Anonymous

In the Evening of his Days

A Study of Mister Gladstone in Retirement

Anonymous

In the Evening of his Days
A Study of Mister Gladstone in Retirement

ISBN/EAN: 9783337418281

Printed in Europe, USA, Canada, Australia, Japan

Cover: Foto ©berggeist007 / pixelio.de

More available books at **www.hansebooks.com**

IN THE EVENING OF
HIS DAYS

A STUDY OF
MR. GLADSTONE IN RETIREMENT,
WITH SOME ACCOUNT OF
ST. DEINIOL'S LIBRARY AND HOSTEL

London:
"WESTMINSTER GAZETTE"
1896
(*All Rights reserved*)

" Change of labour is, to a great extent, the healthiest form of recreation."—THE IMPREGNABLE ROCK OF HOLY SCRIPTURE.

" It has been obvious to us all that they were blest with the possession of the secret of reconciling the discharge of incessant and wearing public duty with the cultivation of the inner and domestic life. The attachment that binds together wife and husband was known to be, in their case, and to have been from the first, of an unusual force . . . . on the one side, such love as is rare, even in the annals of the love of woman; on the other, such service can hardly find a parallel."—ADDRESS TO THE LANCASHIRE AND CHESHIRE MECHANICS' INSTITUTE, 1862.

" Get knowledge all you can, and the more you get the more you breathe upon its nearer heights their invigorating air and enjoy the widening prospect, the more you will know and feel how small is the elevation you have reached in comparison with the immeasurable altitudes that yet remain unscaled. Be thorough in all you do, and remember that, though ignorance often may be innocent, pretension is always despicable. 'Quit you like men,' be strong, and the exercise of your strength to-day will give you more strength to-morrow. Work onwards, and work upwards; and may the blessing of the Most High soothe your cares, clear your vision, and crown your labours with reward."—ADDRESS AS LORD RECTOR OF GLASGOW UNIVERSITY, 1879.

PREFACE

TWENTY-ONE years ago Mr. Gladstone retired from the leadership of the Liberal Party.

> At the age of sixty-five—[he wrote]—and after forty-two years of laborious public life, I think myself entitled to retire on the present opportunity. This retirement is dictated to me by my personal views as to the best method of spending the closing years of my life.

But men propose, the fates dispose; and the retirement which, in 1875, was intended to be final turned out to be the prelude to one of the most active, as it was also one of the most successful, periods in Mr. Gladstone's public life. What political labours he discharged during the two succeeding decades, through what storm and stress he was yet to pass, is known to everybody; and the recital does not in any way come within the scope of these pages. At last the end of Mr. Gladstone's public life arrived; it had been extended to an age greater than that at which any English statesman had ever conducted the Government of his country; and though a fond belief in the Old Man's political immortality might still survive, and though appeals might still be addressed to him to " speak once again and wake a world supine," yet the limit of his public course was now finally attained, and he was free, in the late evening of his days, to satisfy " his personal views as to the best method of spending the closing years of his life."

It is a study of Mr. Gladstone in Retirement that is the object of this little book. Some portion of it has already appeared in the columns of *The Westminster Gazette*. The rest is now published for the first time; and in the preparation of the book, the authoress (Miss Hulda Friederichs) has had the advantage, which the Editor most gratefully acknowledges, of the advice and assistance of some of those who are nearest and dearest to Mr. Gladstone. All the world knows the Statesman. In these pages it is the Man—and, may we not add, the Christian ?—with whom we are concerned. The revelation must necessarily be less complete, less unreserved, than in the case of a public career. Yet the pages which follow will be found to contain much information never before brought together, and will suffice in the hands of any sympathetic reader to present an intelligible and convincing picture of Mr. Gladstone in the Evening of His Days.

In addition to the chapters descriptive of his daily life, it has been our privilege to obtain the first full and authentic account of St. Deiniol's Library and Hostel. These chapters have been prepared with the aid of facilities most kindly afforded us at headquarters. Everyone will be interested in learning the facts about a scheme of public utility which has been in Mr. Gladstone's mind for many years, and the realisation of which has been occupying much of his time since his retirement. But we venture specially to commend this account of Mr. Gladstone's Foundation to all clerical readers, as St. Deiniol's is specially designed for the benefit of Clergymen and Students.*

---

* Mr. Gladstone, having read the chapter dealing with St. Deiniol's, writes to us as follows :—" You are quite at liberty to state, on my behalf, that I think the article has been drawn with much ability and care ; and that, notwithstanding the obvious difficulties of describing an institution still in its embryo, or at any rate not advanced beyond its cradle, it conveys a wonderfully just idea of what I trust may prove to be an useful place."

Of the significance, as a whole, of this study of a great man in retirement, it would be superfluous to speak. The story will signally fail of its purpose if it do not carry its own moral with it. We can best conclude these introductory remarks by applying to the subject of the following pages, in his old age, some words which he applied a generation ago to others:—

> In the sphere of common experience, we see some human beings live and die, and furnish by their life no special lessons visible to man, but only that general teaching, in elementary and simple forms, which is derivable from every particle of human histories. Others there have been who, from the time when their young lives first, as it were, peeped over the horizon, seemed at once to—
>
> "Flame in the forehead of the morning sky"
>
> —whose lengthening years have been but one growing splendour and who at last—
>
> ". . . leave a lofty name,
> A light, a landmark, on the cliffs of fame."
>
> —*Place of Ancient Greece in the Providential Order of the World.*—ADDRESS AT EDINBURGH UNIVERSITY, 1865.

"WESTMINSTER GAZETTE" OFFICE,
January 30th, 1896.

CONTENTS

			PAGE
PREFACE	. . . .		vii
CHAPTER	I.	IN HAWARDEN PARK	3
,,	II.	MR. GLADSTONE AT CHURCH . . .	15
,,	III.	A MORNING WITH MR. GLADSTONE IN HIS LIBRARY	32
,,	IV.	MR. GLADSTONE AT THE LUNCHEON TABLE	46
,,	V.	MRS. GLADSTONE	53
,,	VI.	THE HOLIDAYS OF AN OCTOGENARIAN .	73
,,	VII.	WITH MR. GLADSTONE AT SEA . .	86
,,	VIII.	MR. GLADSTONE'S GIFT TO THE NATION	93
,,	IX.	ST. DEINIOL'S LIBRARY	109
,,	X.	ST. DEINIOL'S HOSTEL . .	134
,,	XI.	MR. GLADSTONE'S HOME-LIFE	144

ILLUSTRATIONS

	PAGE
THE RIGHT HON. W. E. GLADSTONE	*Frontispiece*
THE PARK GATE, HAWARDEN	2
MR. GLADSTONE'S STICKS AND AXES	4
A GLIMPSE OF THE CASTLE FROM THE PARK	6
THE OLD CASTLE, HAWARDEN	9
DOROTHY'S DOVECOTE	13
ST. DEINIOL'S CHURCH, HAWARDEN	17
DOROTHY DREW AND PETZ	19
THE LAKE, HAWARDEN PARK	21
WATERFALL IN HAWARDEN PARK	24
MR. GLADSTONE LISTENING TO THE SERMON	29
HAWARDEN CASTLE, FRONT VIEW	35
PRINCIPAL STAIRCASE, HAWARDEN CASTLE	38
MR. GLADSTONE IN HIS LIBRARY, READING	40
MR. GLADSTONE'S LIBRARY, HAWARDEN CASTLE	43
THE DINING-ROOM, HAWARDEN CASTLE	48
MR. GLADSTONE'S BEDROOM	59

ILLUSTRATIONS

	PAGE
Dining-room in the Orphanage, Hawarden	64
Staircase in the Orphanage, Hawarden	66
Dorothy Drew's Domain: The Night Nursery at Hawarden Castle	67
Dorothy Drew's Domain: The Day Nursery at Hawarden Castle	71
The Morning Room, Hawarden Castle	77
St. Deiniol's Library	99
View over the Dee from the Library	101
A Woodland Hollow in the Park	105
Summer House, and View from St. Deiniol's Library	115
St. Deiniol's Library: The Divinity Room	119
St. Deiniol's Library: The Warden's Room	122
The Rev. H. Drew, Warden of St. Deiniol's	126
The Rectory, Hawarden	128
St. Deiniol's Hostel	135
St. Deiniol's Hostel: The Common Room	138
St. Deiniol's Hostel: The Dining Room	139
St. Deiniol's Hostel: The Prayer Room	141

THE PARK GATE, HAWARDEN.

CHAPTER I.

IN HAWARDEN PARK.

Hawarden as a Tourist Haunt—From Chester to Hawarden Village—The Walk to Hawarden Church—The View from Mr. Gladstone's Study Windows—The Little Lady of the Doves—A Clearing in the Woods—The Gladstone Family as Wood-cutters.

ALL during the summer Hawarden Parish Church is a centre of attraction to the tourist, both American and British. Travellers, on the way to or from Ireland, time themselves so as to arrive at Chester on Friday or Saturday. The American traveller, whatever day he may land at Liverpool, makes a point of staying over the first week's end; and the Britain-enamoured Briton, taking his holiday in the interesting regions of his own country, does the same. Chester, with its picturesque streets, its gaunt remnants of gigantic Roman walls, its red cathedral with crumbling stones, and tattered flags and general look of airy emptiness, is, for the nonce, only taken by the way. The tourist, as in duty bound, goes and inspects everything he is told to inspect, but he glances at it all, as you glance over the preface of an epoch-making new book. The real interest is yet to come. To what "real interest" is Chester the preface, as it were? The next Sunday morning brings the explanation.

The tourist breakfasts early, and then goes off towards the station with an appearance of expectation and determination about him, utterly out of keeping with the Sunday air which seems to steal from "pious Wales" into this delightful corner of Cheshire. He takes his ticket to Broughton Hall or Sandycroft, unless it rains (when he goes by another line straight on to Hawarden station), for, in order to enjoy the day's pleasures to the full, he takes *en route* for Hawarden village a long and lovely country walk, where sea-breezes play around him, while he

MR. GLADSTONE'S STICKS AND AXES.

A GLIMPSE OF THE CASTLE FROM THE PARK.

walks under ancient oaks and elms and beeches; where the sunlight lies broad and golden on a hundred fields; where the sands of Dee, and hills with white and rosy lights upon them, close the wide view to the right; and where, straight before him, rise in dreamy grandeur the spurs of the Welsh hills, reminding him, as does no other range of hills in England, of the first view of the Alps when Basle lies behind and Lucerne in front.

Then there is Hawarden village on the hilltop. Clean and quiet, and just a little sleepy, it looks on a Sunday morning, when the blue smoke rising from the comfortable cottages seems to be the only thing that moves for several morning hours. You have not been in time, you tourist folk, to hear the early bells, and see the small stream of villagers who regularly go into the wrought-iron gates leading to the graveyard and to the parish church before eight a.m. That stream runs steadily each Sunday morning from the village to St. Deiniol's Church, and on the spot where the pigeons, with silvery white wings, flutter around the heavy ivy-covered gates leading into the grounds of Hawarden Castle another human stream flows into that of the village. For there is never a Sunday in the year, while "the family" are at home, on which some members at least of Mr. Gladstone's home circle do not go to the early service. Up to a very recent date one figure was never missing in the party coming through the Castle gates at this hour. Whether the sun shone or the rain fell, whether the storm tossed about the branches of the trees in which he takes such pride and pleasure, or whether from the plains below there rose the thick, white, famous (or is it infamous ?) Welsh mist, Mr. Gladstone himself began his day invariably by paying his tribute to his God in the services of the Church which he believes will always be faithful to her trust, and to which he has adhered all the days of his life.

There is a very beautiful private walk of somewhat over half a mile from Hawarden Castle to the village church, and along that walk Mr. Gladstone has walked innumerable times, bound for St. Deiniol's Church. This walk leads upwards first, past a magnificent sloping lawn, where once stood many of the finest trees on this finely timbered estate. Even now there is a cluster of beeches of such exquisite beauty that, as you look up the long, slender, graceful stems, you are reminded at once, if you happen to have seen them, of those "poems in stone "—the pillars in Cologne Cathedral. Mr. Gladstone, as he sits in his private library, his "Temple of Peace," at that one of the three desks which is known as the "literary" table, looks out on this unique group of slender trees whose heads are up in the blue, higher than the turrets of the Castle.

On the right, as you ascend the path that passes close by the ruins of the old Castle, you observe an innovation. The peacocks and the pheasants have

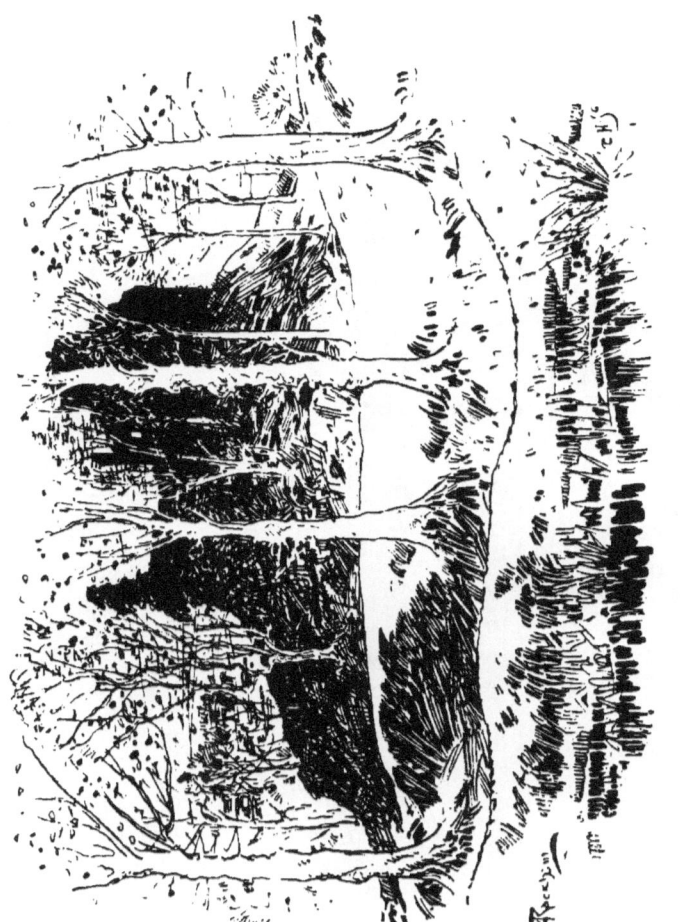

THE OLD CASTLE, HAWARDEN.

long been at home among all this secluded greenery; but there, against the old wall leading to the kitchen garden, is a brand new pigeon-house, with strutting, cooing white inhabitants. A little child has set her toy into the midst of the dignified old park—little Dorothy Drew, Mr. Gladstone's grand-daughter, is the Lady of the Doves. When, on her little bare feet, she descends from the heights of her nursery, at the top of the Castle, to take her subjects for a walk, she marches towards the pigeon-house to take object-lessons in natural history; for of book-learning this lady of five summers knows nothing as yet, her parents holding with the Ruskinian principle that it injures a child's imagination if it be able to read stories too soon. You remember Mr. Ruskin again as you walk along the same path. An oak of enormous size (and Heaven knows what age!) stretches its fantastic branches over you. In Mr. Ruskin's opinion, this tree, to which a century is but a short period of existence, is as near the perfection of a tree as you find anywhere in this country. The lawn itself, sloping towards the Castle and the walled-in Italian garden, was for many years a great delight to Mr. Gladstone. But the storms have played havoc with many of the finest trees on it; and now, though to others who did not know it before some of the giants fell, it looks still a remarkably fine slope, he shakes his head at it, and observes that "it looks like a field."

Through a small Gothic gate you reach the private path in the open park, a green, secluded walk full of bird-life all the year round, and with here and there a fine view of the lower parts of the grounds. On the border of the enclosure, through which this path leads to the gates, a graceful building in black and white wood, not unlike an ornamental Swiss châlet, reflects the sunlight in its large windows. Miss Glynne, the daughter of the late Rector of Hawarden, lives in this house, in the building of which there was an interesting period. For when the spot was decided upon, where the house was to be erected, the whole of the Gladstone family set to work to clear away the trees and the dense underwood with which the site was covered. And they did not play at doing this, but worked as hard and as steadily as might a colony of settlers somewhere in Canadian backwoods, where domestic life could only commence after the ground had been cleared, the log-hut built, and the home-made furniture established. Every member of the household helped. Mr. Gladstone and his sons with their axes felled the trees; and Mrs. Gladstone and her daughters cut off branches, removed underwood, and gave a helping hand wherever they could. Nor did they ever leave off till they were thoroughly tired out, and could look back upon a fair day's work.

DOROTHY'S DOVECOTE.

CHAPTER II.

MR. GLADSTONE AT CHURCH.

Hawarden Churchyard—Mr. Gladstone as Church-goer—His Solitary Walk to Early Service—His Reading of the Lessons—The Old Man in His Pew—"The Day is far Spent"—Listening to the Sermon—Home to Work, "for the Night Cometh."

WHEN once the park gates have been reached it is but a few steps to St. Deiniol's Church. The entrance to the graveyard surrounding it is charming. Square and unbeautiful stone slabs cover the graves to the right and to the left; but among the stones, wherever they could find a foothold in an inch or two of soil, green things and flowers have sprung up, and a fine old yew tree casts its black shadow on the path, up which the Grand Old Man has walked perhaps more often than any of the parishioners. Mr. Gladstone enters by a door in the south side of the aisle, and quietly sits down at the end of the family pew on the right-hand side of the long chancel. And there, even now, as for many years past, the tourist may gaze his fill upon "the squire," who happens to be also one of England's greatest statesmen. Mr. Gladstone believes in regularity in all things, and especially in attendance at church. Nor does he deem it sufficient for anyone to go once on a Sunday to a service, and then stay at home with a calm conscience. As he puts it himself, he thinks nothing of "one-ers."

There is probably not a day in the year—there certainly is not one during spring, summer, and autumn—on which at least one party of tourist-pilgrims does not come to do homage at St. Deiniol's Church, Hawarden. The patron saint, whose statue stands in a niche over the door of the church, is not, however, the chief recipient of this homage. Nor do the crowds which pack the church Sunday after Sunday at each service assemble because St. Deiniol attracts them more than any other saint, or because the service itself differs in some striking way from the ordinary Church of England service. No; the centre of attraction is elsewhere. The eyes of all those who come to "view," or to attend a service in, Hawarden Parish Church wander immediately to the lectern. For at it, for many years past, has stood, whenever he was at his home, the venerable figure of Mr. Gladstone, reading the

Lessons regularly at early morning service during the week, and often at the Sunday services as well. Since the recent operation on his eye he has not stood at the lectern, but he is still a regular attendant at the church.

ST. DEINIOL'S CHURCH, HAWARDEN.

The body of St. Deiniol's Church is old, and of pale grey stone, in simplest Gothic style. Low oak seats come close up to pulpit, reading-desk, and organ. But the chancel, evidently of later date, forms a magnificent contrast. The red stone of which it is built throws a pale, warm, terra-cotta light over the pews and around the altar, and softens even the colours of the painted altar window. Here and there in this dusk you notice the metallic gleams where the light strikes against the crown-shaped brass chandeliers. Then, at the further end, a beautiful altar, covered with cloths of sage green and ruby red; and again, darts of strong yellow light from gold embroideries and from the altar-cross that stands amid offerings of white flowers. Truly this church, on the edge of the hill, in the centre of the graveyard in which "old-fashioned" flowers bloom nearly all the year among the grey stones—this village church, with its associations of the private life of a great man, is a place well worth visiting.

Less than two years ago there was still no more regular worshipper at the eight o'clock morning service at Hawarden Parish Church than Mr. Gladstone; and, though he was over eighty years of age then, he refused absolutely to take even a cup of tea before he went. Sometimes a guest, staying at the Castle, would be up in time to join his host on this early walk. But this could not be done. For though Mr. Gladstone is the most graciously courteous of hosts, with a courteousness which reminds you of the days of Addison and Sir Roger, he would share his early morning walk to church with no one. Not even Petz, the little black Pomeranian, who is otherwise his master's inseparable companion on all walks about Hawarden, ever so much as suggested to walk to church at 7.45 a.m. In silent thought, and with brisk step, Mr. Gladstone walked along the green path and entered the church alone, to follow the service with undivided attention, and, very often, to assist the officiating clergyman by reading the Lesson of the day. But at the conclusion of the service

DOROTHY DREW AND PETZ.
(*From a photograph by Messrs. Elliott & Fry, Baker-street, W.*)

he threw off his reserve, joined his friends, and was ready—nay, eager—for the discussion of any subject in heaven or upon earth which might happen to be the *spécialité* of a guest. More than one of those who thus have walked back to breakfast with Mr. Gladstone have told me that never at any time, under any circumstances, have they seen Mr. Gladstone more keenly alive, more brilliant, more rapid in seizing new points, more pertinacious in defending an opinion, more genial and fascinating, than during these morning walks. But attendance at early services is now a thing of the past; since his sight has been affected Mr. Gladstone has been obliged to be very careful concerning the use of his eyes, and when an attack of influenza somewhat weakened his general health, he gave up this habit of many years. This tribute to old age was not an easy one, but Mr. Gladstone is far too wise, too reasonable, to go against the word of command from the superior authority, the oculist, who joins with the physician in saying that Mr. Gladstone is a perfect patient.

With regard to Mr. Gladstone as reader of the Lessons at church, a good deal of sympathy has been wasted by those who imagine that he took this part of the service "because he liked it." "Poor Mr. Gladstone," they say, "he must feel it very much when he goes to church that he can no longer take his part, as he used to do, by reading the Lessons of the day." Mr. Gladstone *does* feel it, but he feels it as a relief and not as a deprivation. As a matter of fact, Hawarden parish is large and scattered, and the staff of clergy was, until recently, hardly sufficient for the needs of the flock. Thus the work, even when divided and regulated in the most careful manner, was very heavy, and it was only

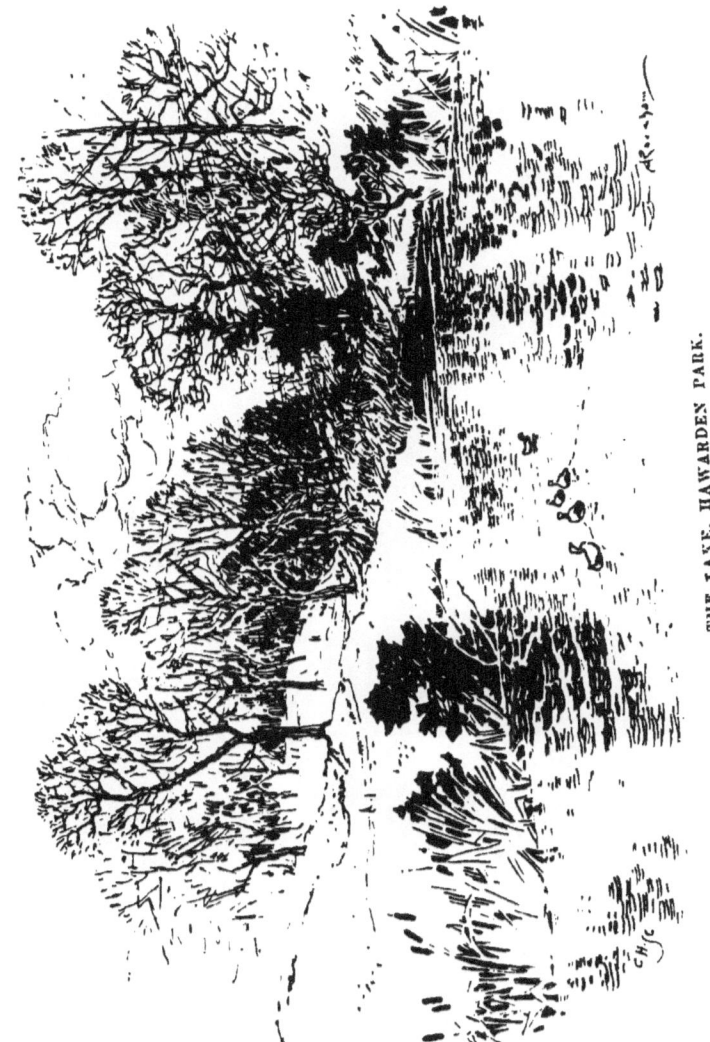

THE LAKE, HAWARDEN PARK.

in order to assist the officiating clergyman, but in no way because he himself preferred to take an active part in the service, that Mr. Gladstone assisted by reading the Lessons. Now that the Rector has a large number of assistants, the work can more easily be done without a lay-reader. Therefore Mr. Gladstone has retired from this self-imposed task, not regretfully, but well content, thinking that he has, at eighty-six, earned rest so far as to be exempted from the labour of helping the clergy, especially since he got out of the habit. On Sunday evenings Mr. Gladstone still reads Prayers at home.

The Rectory, where his son, the Rev. Stephen Gladstone, has his home, forms, as it were, another link in the chain of buildings formed by the Library, the Hostel, and the Church. It is a fine old red brick mansion, overgrown with roses and magnolias, and the grounds in which it stands are very beautiful, with lawns and giant trees and copses, winding walks, and banks, and parterres of flowers. Ever since he first came to Hawarden Mr. Gladstone has made the Rectory grounds one of his favourite haunts. At Easter, when he always stays at the Rectory, he spends long hours in the grounds as often as the weather is at all favourable. He has perhaps read and thought as much under these trees as anywhere else, and still, as life's eventide draws near, he holds the place most dear.

Hawarden Park, at the time when the materials for these chapters were gathered, stood in all the glory of October's golden crown. The chestnut trees seemed to radiate yellow light, the beeches and oaks were billows of coppery red, the roses against the Castle walls, and in the quaint Italian gardens, were still in bloom; and amidst these very silent, very beautiful scenes Mr. Gladstone leads his simple life among his own friends, and with a constant stream of visitors, not one of whom leaves Hawarden Castle without being more than ever impressed with Mr. Gladstone's marvellous vitality.

In the wintry days, the daily pilgrimage of travellers and tourists is discontinued to a great extent, for it is only on fine days that Mr. Gladstone is likely to be met in his old haunts about the estate or country lanes. When the days are bleak—as, unfortunately, they mostly are between November and April—he is far too happily employed at home to venture out into the raw, grey winter day. We shall presently return to the subject of his home occupation—to the ten-hour working day of this veteran of eighty-six. For the present it is interesting to return to Mr. Gladstone as a church-goer. The Hawarden bells which seem to have a peculiarly harmonious ring in them as their voices sing over the wide plains on the borders of musical Wales, never call Mr. Gladstone in vain, so long as the weather is not too cruel, when

WATERFALL IN HAWARDEN PARK.

they ring for service at eleven on Sunday mornings. You see him briskly walking along the private road, perchance with a family group, or with a guest; perchance alone, a little in advance of the rest; and, unless the weather be particularly inclement, he wears no out-door clothing but the short shoulder cape in which the cartoonist has so often depicted him. Even now, when his years are approaching fourscore-and-ten, he clings to the habits of his early manhood, and one of these is a rooted dislike to any great coat which comes down to the knees, and impedes the freedom of the limbs. As you see him hastening along in the distance he seems not one whit changed since the days when thus he strode across St. James's Park to take the harmless, necessary constitutional between long hours of work. And you rejoice, and think for the moment that perhaps they are not so very foolish who, even at this hour, refuse to believe that Mr. Gladstone has retired from public life to return no more.

But when, ten minutes later, you see him again, in the corner of the pew, dimly visible in the subdued light of the chancel, you are bound to change, however reluctantly, your recent opinion. It is, indeed, a very old man who sits there, with bent back and pale, wrinkled face, holding his book close to his eyes, apparently unable to find or to follow the text. A stray sunbeam, white as midwinter sunbeams mostly are, falls upon him from behind. The ivory white of the bald, magnificent head gleams in the light, and the few silvery hairs round the base of the head form a striking contrast to the glittering gilt edges of the book he holds. "The day is far spent." And you turn sadly from the contemplation of the silent, quiet figure.

The service goes on, and presently the preacher ascends the pulpit. Then from his retired chancel seat Mr. Gladstone rises and comes to sit opposite, close to the lectern, where, on the wings of an eagle, lies the Book of Life from which he has so often read to the congregation assembled here. And now again, you see that you were altogether wrong in imagining the Mr. Gladstone of a few years past to have entirely vanished. For you have but to concentrate your sight on the old man on the wooden seat before you, and to forget his present surroundings, and you have Mr. Gladstone, in one of his most characteristic House of Commons attitudes. His head is slightly bent; his face, save for an occasional movement of the eyelids, perfectly motionless; he bends forward, folds his hands over his knee, and listens with unflagging attention. Now and then, as a passage in the sermon strikes him in a particular manner, he holds his hand in the well-remembered way to his ear, that not a sound may escape him. The attention of other members of the congregation may momentarily flag, or their eyes may wander, but Mr. Gladstone is all

MR. GLADSTONE LISTENING TO THE SERMON.

attention from beginning to end of the sermon and service. What an ordeal, by the way, for a newly-fledged young priest to preach to that listener on the wooden seat under the lectern! What an ordeal; and what a record to look back upon in after years!

After church you see the Old Man once more buoyantly walking along under his trees; he cannot afford to linger, for there is yet a good hour before luncheon in which to get on with the "Analogy," the stiffest book in the world to some other mortals; the source of endless fascination to Mr. Gladstone, the theologian and fervent admirer of Bishop Butler.

CHAPTER III.

A MORNING WITH MR. GLADSTONE IN HIS LIBRARY.

The Vitality of an Octogenarian—Mr. Gladstone's Breakfast—"If you want a thing done well, do it yourself!"—"The Temple of Peace"—How He Arranges His Books—His "Political" and his "Literary" Desks—His Powers of Concentration—His Pet Dog "Petz."

THERE is not one of the constant stream of guests at Hawarden Castle who does not go away wondering more and more, at the end of each succeeding visit, at the vitality of the master of the mansion, the oldest man of the circle gathered around him—the oldest and also the most vigorous. "It is marvellous," Mr. Gladstone's friends tell you; and those who look gravely and seriously at old age, add thoughtfully, "and it is a most beautiful thing to see a long life ending in such perfect happiness and peace."

But not the most vivid account can give a true

idea of Mr. Gladstone at the age of eighty-six; of
the strength of his individuality; the keen, rapid
insight into matters; the eager interest he brings to
such subjects as may be new to him in so far that
he has never made a special study of them; the
humour and good-natured raillery with which he
intersperses his occasional observations, and the sense
of constant activity in which there is not the slightest
suspicion of unrest. In repose, at church, he looks
physically feeble; when you see him energetically
walking out of doors, you modify your opinion
concerning the supposed feebleness; but only when
you see him at home is it possible to realise his
marvellous vitality, which is far above that of the
majority of men half his age.

Mr. Gladstone begins his day's work at eight o'clock
every morning. That is to say, he begins to read.
Having taken breakfast in bed—a luxury which he has
only quite lately allowed those about him to impose
on him—he does not rise at once, but reads for two
hours, in preparation for the work he has set himself
for the day. At ten a.m. he gets up, and is dressed in
less than half an hour. The idea of a valet assisting
him never enters his mind; he would scorn to entertain
it for a moment. Not even will he suffer anyone to
assist him in the most important part of his packing
when he goes away from home. On the day preceding
a departure for, let us say, a lengthy stay at Biarritz

D

or a cruise in the *Tantallon Castle*, every bed- and dressing-room at Hawarden Castle, where the business of packing is going on, is in the state of confusion which generally reigns in the private rooms of intending travellers just before their departure. But of all this there is nothing in Mr. Gladstone's rooms. Every pin is in its right place, every book on its shelf, and all the books, the writing materials, the papers—in fact, everything he may require during his absence except his small portmanteau of clothes and toilet necessaries—is neatly packed by the owner himself, without the slightest confusion, on the principle "if you want a thing done well, do it yourself!"

At Hawarden, Mr. Gladstone, coming from his bedroom, goes downstairs, and turns at once into his library. No wonder it has been named the "Temple of Peace," for even as you enter it the air of deep peace strikes and falls upon you. It is a library quite unlike the generality of such apartments in country houses, where a stray visitor goes to write his casual notes; whither a man withdraws when he is tired of the society of his kind for a couple of idle hours; or where young couples flock for refuge when large house-parties make all other rooms unsafe for an enjoyable *tête-à-tête*. About the books in the average private library, all bound in accordance with the barbaric notion that backs must look alike, and each shelf must contain only books of the same size, there hangs almost invariably

HAWARDEN CASTLE, FRONT VIEW.

a suggestion of mustiness and dustiness that makes the real lover of books wince or smile, according to temperament. In the Hawarden Castle Library—or, as I was amused to hear it called the other day, "the Castle section of St. Deiniol's Library"*—there is no trace of all this country-house librarism. It is a room which is lived in; which bears, wherever you may turn, the impress of its owner's tastes; which has been beautified with portraits and bas-reliefs of the owner's children and nearest friends, with curios and keepsakes; and which, for all its rest and quiet, is full of the spirit of work. The books on the shelves are arranged not according to size, but to subject; they are not stupidly pressed into the same uninteresting state livery of calf or morocco, but they have all preserved their individuality by retaining the clothes in which the publisher dressed them before they went out into the service of the world.

By an ingenious invention, Mr. Gladstone has arranged all his books in such a manner that not one of them is hidden behind others, and that he can see the titles of them almost at a glance. That is to say, anyone with unimpaired eyesight can do this. As for Mr. Gladstone himself, he said a few days ago, with a half

---

* The visitor to St. D.'s Library will observe among the notices on the board that "Books at present in the Castle section of the Library may be obtained on application."

impatient, half-humorous shrug of the shoulders: "Oh, no! to me the arrangement is not much good, now that

PRINCIPAL STAIRCASE, HAWARDEN CASTLE.

my eyes are bad. I can't see anything when I look rapidly along the backs of the books. I must focus each one separately." But he lingers with quiet enjoyment over a talk about his books, and likes to hear you praise the arrangement of the shelves, which is the same at St. Deiniol's Library and all round the "Temple of Peace." One of Mr. Gladstone's *spécialités*, by the way, is his wonderful knowledge of the Bible, which those who know him best declare never to have seen surpassed, and rarely equalled, even by clergymen.

Three writing-tables stand in the Hawarden Castle Library, and not one of them is for show. At one table Mrs. Gladstone writes many of her letters; the second, standing between two windows, is Mr. Gladstone's political desk. It looks now too neat, too tidy by far for the taste of any Gladstonian politician; the blotting pad has an unused air about it; the pen-wipers, racks, knives, and other appurtenances stand about with a mathematical precision which means want of use, and there is neither note nor leaflet to be seen. The political desk, at all events for the present, must be numbered among the unemployed. But turn to the third, the literary writing-table, in the niche by the window, which is the most peaceful corner in that place of peace: you may there, at any time during the day, see a sight that will make your heart rejoice, and that will linger in your mind while pictures linger there at all.

The whole desk is covered with reference books of all shapes and sizes. They form a wall round three sides, and there is only an opening in the front of the desk just wide enough to allow Mr. Gladstone to place the sheets of foolscap paper on which he writes. There, at this desk, the G.O.M. sits all day long, with very short intervals, absorbed in "Bishop Butler," the man

MR. GLADSTONE IN HIS LIBRARY, READING.
(From the Portrait by Mr. J. McLure Hamilton, now in the Luxembourg Gallery).

whom he gives a place among the greatest of men that have ever lived. The number of volumes required by Mr. Gladstone for reference is almost endless; he has often to search through one bulky tome after the other for some chance paragraph that may help him to elucidate a dark place in the "Analogy." There he sits, absorbed in his work: now writing rapidly and eagerly for some minutes; now throwing down his pen, and dipping deep into one of the books with which he has surrounded himself. Sometimes a specially-favoured artist has been admitted to take Mr. Gladstone's portrait, but it is on express condition that the "sitter" shall not in any way be distracted from his work. It was under such conditions that Mr. McClure Hamilton's portrait (now in the Luxembourg) was taken. Not once does Mr. Gladstone's attention flag, though you may perchance be sauntering about the room in company with a member of the family, or be privileged to have a chat with Mrs. Gladstone as, with loving pride in all her husband's doings, she points out to you this, that, or the other of the objects in the library to which attaches a special interest for him. Not that Mr. Gladstone is altogether indifferent to violent interruptions. In fact he objects to them strongly, and he does not hesitate to say so. But when it happens that he himself has invited you to "come into my room," he forgives you if you appear while he is at work, and works on undisturbed by your presence or your voice, with an absorption in his occupation which is nothing short of marvellous. Whatever the work may be which he has in hand, it takes hold of him so entirely that he has to be roused from it as most people are roused from sleep. Mr. Gladstone does not as a rule care to talk of his idiosyncrasies, since he holds the sincere opinion that he is very much like the rest of the world. But he acknowledges that his faculty of concentration is his special gift, the one by which he is distinguished from other people.

Another reason why Mr. Gladstone gets through so astounding an amount of work is his extraordinary habit of using up odds and ends of time. One day, not long ago, he was going for a drive into Chester after luncheon. His pudding was very hot, so he went away from table, changed his clothes, got ready for the drive, and came back and finished his meal, thus saving the ten minutes during which his pudding cooled! It may here be mentioned, in connection with the drives to Chester, that on the day, a few months ago, when he drove in for the purpose of making his powerful Armenian speech, Mr. Gladstone had been absorbed all the morning in Butler, and the speech was made without any special preparation.

From ten a.m. to luncheon-time Mr. Gladstone spends regularly in the library, hard at work. But when the gong sounds, he puts down his pen, for promptness in everything has become a second nature to him, and he is probably in the dining-room before anyone else arrives. Petz alone, if that little creature can arrange it so, is as prompt as his master, for his two ruling passions—love for his master and love for his dinner—come into play at the sound of the luncheon gong. The dog's small black muzzle is

MR. GLADSTONE'S LIBRARY, HAWARDEN CASTLE.

turning whiter and whiter, and his eyes are growing just a little dim. But his spirit is proud and high, and if, after having done justice to a plentiful meal, he espies an enemy in the shape of a cat in the garden in front of the dining-room, he treads the cushioned seats in the oriel window with stern step and angry bark, and demands instant admission, through the window, into the garden, that he may purge the neighbourhood of such " vermin."

CHAPTER IV.

MR. GLADSTONE AT THE LUNCHEON TABLE.

About *The Westminster Gazette*—Tales of the "Thirties" and the "Forties"—Talker and Listener—Mr. Gladstone's Hearing—His Light Reading—A Note on his Edition of Bishop Butler.

AFTER five or six hours' literary work on so heavy a subject as the editing of Bishop Butler's work, even a young man might be forgiven if he came to the gathering at meal time in a somewhat silent mood. Not so Mr. Gladstone. The "Analogy" has for the moment ceased to exist; there are an endless number of other subjects which are of interest, and a change of occupation, whether it be mental or physical, is as good as a rest. Hence, as Mr. Gladstone sits at the head of the table, his face is the most animated of all. The light pours fully upon him, and you are anew compelled to admiration of the fine head, the eagle eye, the rapid change of expression, which, however, never drives away the underlying look of large, dignified benevolence.

In two minutes the conversation is in full flow. One subject only is now mostly forbidden—that is, controversial politics. On anything else your host is ready to talk, and, whatever the subject be, it is sure to become interesting. Take journalism. The name of *The Westminster Gazette* happens to come up—and, we say it with a soothing sense of gentle self-satisfaction, is dwelt on by Mr. Gladstone in a most flattering manner. How is an evening paper produced? When is it written? How does the work compare with that on a morning paper? Is *The Westminster Gazette* aware that years ago there was a Roman Catholic paper of the same name? It is interesting to note that this MS. was edited by Mr. Purcell, Cardinal Manning's biographer from 1866 to 1879. The Turkish Question cropped up, just as it does to-day, and the editor had anti-Turkish opinions, and said so in the first *Westminster Gazette*. All the Catholics were "Turks," except Lady Georgiana Fullerton and a few other enlightened people. The result was, the paper died of a disease called "Turk."

Then the talk turns to the new peers, and very wittingly, and evidently with a keen sense of enjoying his reminiscences, Mr. Gladstone dips again into the past of sixty years ago, and remembers the ancestors of one of the peers, and their doings in the times of their youth. It is all told as cheerily as a schoolboy will tell

THE DINING-ROOM, HAWARDEN CASTLE.

you of the good fun he has had last term with Tom, Dick, and Harry, his intimates or his opponents. Not an incident is forgotten, and as you listen the days of "the thirties and forties" live again in the nineties, and are as vivid as is yesterday. All at once there is a pause, and a twinkle comes into the eyes of the host as he turns round and says with assumed gravity : " I could tell you ever so much more, but I won't. I don't want you to get me into a libel action."

A moment later you see a man who differs widely from the brilliant talker. Mrs. Gladstone has not been well; it was not thought wise that she should leave her room before luncheon. Now, Mrs. Drew, who is the stay and staff, together with her husband (the Rev. H. Drew), of both her parents in their old age, reports that the patient is coming down into the library, to wait for the carriage which is to take her for her first drive after her illness on this golden winter afternoon. And gravely, and just a little anxiously, Mr. Gladstone now turns his attention to the patient, and asks about her luncheon, etc., etc., and brightens up again as he hears that all is as well as can be.

Books are a subject to which Mr. Gladstone returns with ever new delight, and the books in St. Deiniol's Library are naturally discussed in these days when that institution is about to be made over to trustees— a subject, by the way, to which we shall presently return. Talking of theological books, and seeing that Mr. Gladstone's chief occupation is the editing of a theological work which necessitates the use of an enormous number of reference books, the question naturally suggested itself: " But do you not miss the thousands of books you have sent down into St. Deiniol's Library ?" And the answer, though given in a tone of playful grumbling, is rather pathetic, since it is the exact truth; "Yes, I do miss them. I miss them every day and every hour. When I want a book it is sure to be at St. Deiniol's." But noticing that his friends are rather saddened by this unvarnished expression of the truth concerning the theological books which have gone from the Castle to St. Deiniol's, Mr. Gladstone at once falls in with the suggestion of establishing some modern communication between the two parts of the Library, by means of which a book can be called for at a moment's notice.

Then back to modern journalism once more. It is a marvel how quickly things are done nowadays, Mr. Gladstone thinks; and cross-examines the journalist a little more, and proves himself as good a listener as he is a talker. And here, by the way, it may be observed that the idea is erroneous that Mr. Gladstone's hearing is altogether failing. Sitting by his side, you need not raise your voice in the very least; you need not even be careful to utter your words with the "extra-special" distinctness which it is always wise to assume when talking to a member of that great army of old and elderly people who are slightly deaf but object to being thought so. At the same time, it was one of Mr. Gladstone's reasons for retiring from politics that, during his last Parliament, he never heard the whole of a speech to which he had to reply.

When at last, after a light meal, and a half-hour of

talk, grave and gay, Mr. Gladstone rises from table, he at once goes back to his "Temple of Peace," and settles down to more work, just as eagerly, just as deeply absorbed as before. One very stiff part of the editing of Butler's work, which was recently finished, was the indexing of the forthcoming edition. Even this Mr. Gladstone did entirely himself; and on the evenings of the indexing days he confessed himself in so far beaten by his work that he required an antidote of particularly light reading, and renewed old friendships after dinner by plunging into "Robinson Crusoe" and "The Arabian Nights."

Light literature, a good deal of which is fiction, occupies Mr. Gladstone at the end of every ten-hours' working day after dinner, unless, of course, his guests draw him away from his books. Among the books which he has read with pleasure during the last month or two are "Aunt Anne" and "A Flash of Summer," also Sir W. Hunter's "The Old Missionary." From "Obscure" books, however much they may occupy the table-talk of the day, Mr. Gladstone saves himself, and rather than be racked by the realist's scenic display, he turns back to Miss Edgeworth, and wanders with her through the realms of the long ago.

In a footnote which Mr. Gladstone added to the article on Matthew Arnold, in the December number of *The Nineteenth Century*, he pointed out that both the "Analogy" and the other principal compositions are broken up into short sections for the greater convenience of reference. It is interesting to note that Mr. Gladstone considers this one of the most important and useful parts of his work, which, in itself, would almost justify a new edition. And thus it happens that the New Journalism has been paid the great compliment of having one of its inventions—that of "cross heads" and "short pars."—adopted by the "most Conservative member of the Conservative Party."

CHAPTER V.

MRS. GLADSTONE.

The Family Life of the Gladstones—A Sketch of Mrs. Gladstone in the "Temple of Peace"—The Ideal Wife in Real Life—Mr. Gladstone's Marriage—Mrs. Gladstone as Guardian Angel—Her Charitable " Homes " at Hawarden.

IT is impossible to spend a day, or even an hour, at Hawarden Castle without becoming aware of what is no doubt one of the chief reasons for the absolute peacefulness of Mr. Gladstone's old age. There is that about the whole household which tells you that " the Gladstones " are what is usually called a united family. The majority of families, of the class to which Mr. and Mrs. Gladstone and their children belong, have at first sight the appearance of being genially attached to each other. But in a good many cases these appearances are deceptive, and when the thin veil of good manners is cast aside you find that the apparent attachment is nothing but a hollow mockery. Not so in the family at Hawarden Castle, which is united by an unusually strong bond of affection, drawing the various members more closely together year by year. Thus, for instance, the widow of Mr. W. H. Gladstone, who is living in the village of Hawarden, and whose little son is the heir of the estate, is now even more at one with the Gladstone family than during her husband's life-time. And the one member of the family round whom cling most closely the affections of all is Mr. Gladstone himself. There is no suggestion of "keeping up appearances," no simulated hero-worship, in the great love and admiration which every member of his household gives to him freely and at all times. Mr. Gladstone is, in very truth, a hero in the eyes of all who are nearest and dearest to him. As Mr. George Russell says, at the conclusion of his deeply interesting biography of his former political chief and intimate private friend :—

> There are some people who appear to the best advantage on the distant heights, elevated by intellectual eminence above the range of scrutiny, or shrouded from too close observation by the misty glamour of great station and great affairs. Others excel in the middle distance of official intercourse, and in the friendly but not intimate relations of professional and public life. But the noblest natures are those which are seen at their best in the close communion of the home, and here Mr. Gladstone is pre-eminently attractive. The dignity, the

order, the simplicity, and, above all, the fervent and manly piety, of his daily life, form a spectacle even more impressive than his most magnificent performances in Parliament or on the platform. He is the idol of those who are most closely associated with him, whether by the ties of blood, of friendship, or of duty ; and perhaps it is his highest praise to say that he is not unworthy of the devotion he inspires.

And first and foremost among his devotees at home stands, naturally, Mrs. Gladstone.

It is fifty-seven years since Mrs. Gladstone and her sister, the first wife of George Lord Lyttelton, were married on the same day at St. Deiniol's Church, Hawarden. Sir Francis H. Doyle, who was Mr. Gladstone's best man, celebrated the occasion in a poem called "The Two Sister Brides." In it occur the following lines addressed to Mrs. Gladstone. Reading them, with the experience of those fifty-seven years behind us, we know that the poet was a seer, and predicted exactly the part which "the beautiful Miss Glynne" would play on the stage of the life she was then just entering :—

> High hopes are thine, oh ! eldest flower.
>   Great duties to be greatly done ;
> To soothe, in many a toil-worn hour,
>   The noble heart which thou hast won.
>
> Covet not, then, the rest of those
>   Who sleep through life unknown to fame ;
> Fate grants not passionless repose
>   To her who weds a glorious name.

> He presses on, through calm and storm
>   Unshaken, let what will betide;
> Thou hast an office to perform,
>   To be his answering spirit-bride.
>
> The path appointed for his feet
>   Through deserts wild and rocks may go,
> Where the eye looks in vain to greet
>   The gales, that from the waters blow.
>
> Be thou a balmy breeze to him;
>   A fountain, singing at his side;
> A star, whose light is never dim;
>   A pillar, through the waste to guide.

For more than half a century she has stood by his side, the most devoted and loyal of "partners;" she has triumphed with all her husband's triumphs; his life, indeed, has been her life, and in days of anxiety or ill-health it is Mrs. Gladstone again who has lightened the burden by never-failing devotion. It is so even now. Mr. Gladstone's quiet happiness is reflected is Mrs. Gladstone's bright and still beautiful face; and there is a touching *abandon* in her voice and manner as she points out to you, one after the other, the things in the "Temple of Peace" which are of special interest. In one of the niches her husband sits silently over his paper, quill in hand, and in the niche opposite sits, perchance the graceful white-haired lady, in soft, clinging garments of some creamy material, brightened about the throat by quaint gold ornaments, and wrapped round with dark velvety furs—for she is waiting to drive out after her recent indisposition. "We don't like to talk when he is at work," she says, in an eager whisper and with a beaming face, "but he does not mind it;" and continues to show you the things which she thinks may interest you most.

If ever wife has approached to Solomon's ideal woman it is Mrs. Gladstone, in whom are realised, to the very letter, the characteristics of the woman whose price is "far above rubies."

> The heart of her husband doth safely trust in her. . . . She will do him good and not evil all the days of her life. . . . Her husband is known in the gates. . . . Her children arise up and call her blessed ; her husband also, and he praiseth her. . . .

And in this also has Mrs. Gladstone, for more than half a century, been an ideal wife, that she has done the work of a guardian angel from no other motive than that of love and admiration of her husband. No ambitious thought, either for him or for herself, has ever marred the perfect unselfishness of her loving care; Mr. Gladstone was her husband, and for that reason his life and well-being were dearer to her than all things else, and whether he was Prime Minister of England, or whether he was simply the scholarly private gentleman, was a matter of accident only.

When "the beautiful Miss Glynne," sister of

Sir Stephen Glynne, of Hawarden Castle, first met the young M.P. with the eagle eyes and the expressive, mobile face at a dinner-party, now fifty-eight years ago, a member of the Government of the day said to her: "Mark that young man! He will one day be Prime Minister of England." Before another year had passed Miss Glynne and that young man were married. They had met again in Italy; and not long after the bells of Hawarden Church pealed for two lovely sister-brides on the same day. There are a few old villagers still alive who recount the story of the double wedding, the greatest event in their lowly lives.

And from the day of their marriage to the day when, in his eighty-fifth year, Mrs. Gladstone, though in delicate health, accompanied her husband on that last journey to his Sovereign, when he had to lay down his life's work, which was becoming too hard as the light began rapidly to fail, she has been in every respect an ideal wife.

For herself Mrs. Gladstone never had any ambition but in estimating the work which Mr. Gladstone has done it should not be forgotten that it is largely owing to her that he has been able to get through the mass of work which is associated with his name. She made it her duty from the first to regard her husband's health as superior to all other considerations—a sacred trust committed to her care which was the first of her responsibilities. Blessed herself with a

MR. GLADSTONE'S BEDROOM.

perfect constitution and unbroken health, she has watched over him with the skill of a nurse and the vigilance of a guardian angel. She knows the limits of her own skill to a hair's breadth, and the moment they are passed she calls in the doctor. Nor is it only in the maladies of the body in which she has displayed invaluable qualities. She has carefully kept Mr. Gladstone shielded from all the minor worries of life. Rightly regarding his work for the nation as infinitely more important than anything he had to do for his family, or the small businesses of life, she has taken upon her shoulders all the routine work which could possibly be devolved upon her, and stood between her husband and the little worries and frictions of life with a perfect genius of thoughtful and anticipatory skill. Who, having once seen Mr. and Mrs. Gladstone together, bent on performing some public duty, or rejoicing the heart of private friends with their presence at some social gathering, could ever forget the touching sight of the man and woman who have grown old together in perfect unity of spirit, and who, at life's eventide, are still both following their great ideal of living useful and unselfish lives? Other women younger and stronger than herself, might grow tired, and disappear from the hot, stuffy cage above the House of Commons known as the Ladies' Gallery; but Mrs. Gladstone remained at her post, forgetful of late hours, forgetful of discomfort and weariness, and with eyes only for the great orator below, at the sound of whose voice the House became hushed and attentive. Her anxious heart might flutter as she looked at his pale and tired face, and noticed the first sign of huskiness in the melodious voice. But she was not, therefore, going to add, by irritating though well-meant admonitions and supplications, to his petty cares. No; his health might be all the world to her, but his life, his strength, belonged to the nation, and she could only sit by and watch and see that the innocent elixir of life she had so carefully prepared for him at home should not run short. This fillip—which has become historic, like Prince Bismarck's brandy-and-water in the Reichstag—is a humble compound of sherry and eggs, prepared at home, and by Mrs. Gladstone's own hand.

It is always "a study" to watch Mr. and Mrs. Gladstone in social life. The moment after they have saluted their hosts they are, of course, swallowed up, separately, in a maelström of admiring friends. And while Mr. Gladstone, eager and intense, is soon absorbed in whatever subject may by chance have been broached, you cannot but note that, notwithstanding her smiling, kindly, sympathetic attention to those around her, Mrs. Gladstone is always conscious that hers is the post of the guardian angel over her husband. Her eyes search for him again and again, and when she sees him happily ensconced in some cosy corner, together with a friend, or beaming in his benevolent kindness upon some fair admirer, a look of quiet satisfaction spreads over the mobile face of his wife.

Again, at bazaars, meetings, and similar functions Mrs. Gladstone, long past the threescore years and ten, is always ready to act with or for her husband, to stand for hours at a stall selling his portraits, to sit through wearisome ceremonies or stand in crowded rooms, if so she may help him to aid the cause to which he has devoted his life and his great powers.

Mr. Gladstone is fully sensible of what he owes to his wife, nor has he made any secret of the fact that his continuance in public service was dependent upon the health of his partner in life. Had she broken down, and become an invalid, he would have retired from the service of his country. It would have been impossible, he felt, to carry on the work of the Government and at the same time to have attended to his duty to his wife; nor could he have stood the strain if she, who had been throughout as a ministering spirit, instead of aiding him, became a tax upon his vitality. The self-denial and self-abnegation of Mrs. Gladstone are beyond all praise. It, no doubt, seems very imposing and dazzling to many to be the wife of a Prime Minister, but the wife herself has a somewhat hard time of it. The absorption of a Prime Minister in the work of a nation leaves him very little time for domestic intercourse. Mrs. Gladstone has been known to remark that, when Mr. Gladstone was in office, and in London during the season, it was quite a treat to her to be invited to a friend's house to dinner together with her husband. She always then tried to get seated near to him—" when," she said, " it is at least possible

DINING-ROOM IN THE ORPHANAGE, HAWARDEN.

or me to have some conversation with my husband; otherwise I see nothing of him."

Apart from the fact that whatever concerns or interests her husband is, as a matter of course, of importance and interest to her, Mrs. Gladstone is not a keen politician; but with her usual readiness to be useful in advancing his good cause, she became president of the Women's Liberal Federation at an age when the majority of men and women consider themselves abundantly entitled to throw off all trammels of this kind. Right faithfully she performed her duties to the Federation, till she felt that, on the verge of her eightieth year, she could no longer retain this post.

Over and above her chief interests, which are all that concerns her husband, Mrs. Gladstone has, close to her own home, many of the interests which lie nearest her heart. Nestling close to the imposing walls of the Castle stand two grey stone buildings. One of these is Mrs. Gladstone's Home for Orphan Boys, and in the other a number of impecunious and lonely old ladies have found a harbour of refuge whence they need not again sail forth on the sea of life. More unconventional charitable institutions there probably never have been than these two Homes. Some thirty orphaned lads are housed in one of the quaint old places. You see them marching through the park, their Eton collars glorious in the sun, a band of cheery, red-cheeked boys, each one of whom is the cavalier and friend of Dorothy Drew, who revels in a gambol with " her boys.". Or you see them seated round long tables in their pretty dining-room, doing justice to the best of their capacity to plentiful meals of well-prepared and wholesome food; or, again, they are about in their playground, or working at their garden plots. And everywhere they look *soignés* and sheltered, these children who have drifted friendlessly upon the world.

F

And they receive at the Hawarden Home far more than education, food, and shelter. For if their quaintly

STAIRCASE IN THE ORPHANAGE, HAWARDEN.

beautiful surroundings, in that house of ancient oak, mysterious nooks, and picturesque passages and windows and stairs and attics, do not develop their æsthetic tastes, it must be their own fault. And in the mansion close by the constant nearness of one of the greatest men of the time, the good influence which in manifold

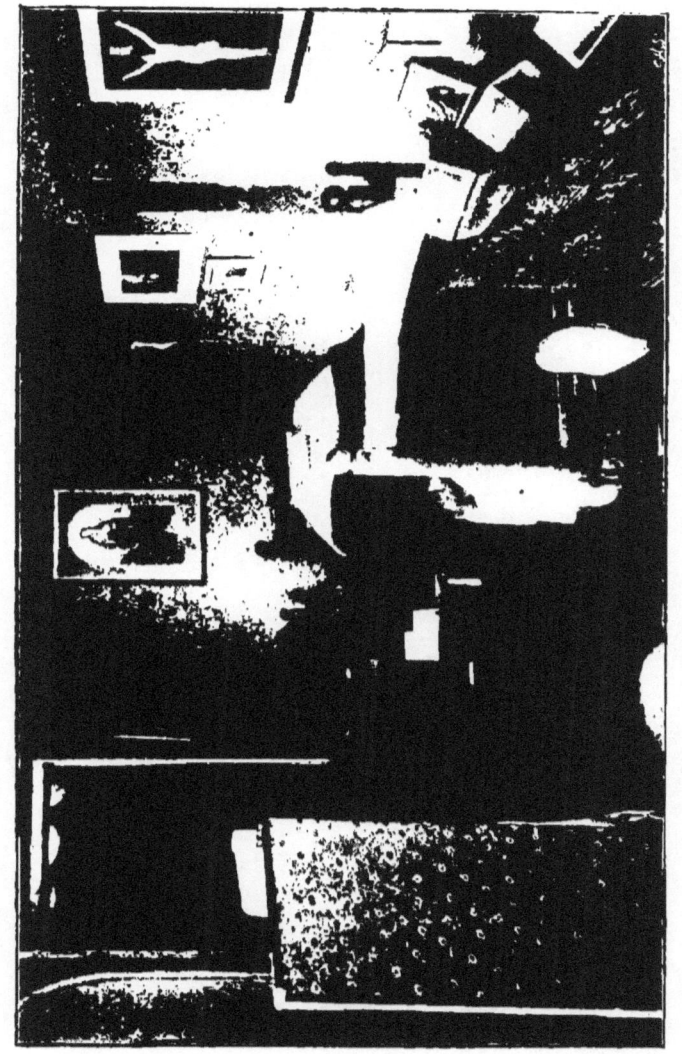

DOROTHY DREW'S DOMAIN: THE NIGHT NURSERY AT HAWARDEN CASTLE.

forms is spread all round by Mr. Gladstone and his family, must needs affect the boys in some way, however subtle and silent, moulding their future, and giving them a far better chance in the race than generally falls to the lot of children who are orphaned and poor.

The old ladies next door to the Orphanage are in their way made just as happy as the boys. To see them in that sunny little house, in their white grandmother caps, their neat and even dainty attire, and surrounded by a hundred things that make for comfort and brightness, is like looking through one of Kate Greenaway's or Randolph Caldecott's picture-books. " Upstairs and downstairs and in the ladies' chamber " it is the same, and after bidding the gentle cronies and the cheery matron good-bye, you seem to see above the doorway of this home the old, old motto: " At evening time it shall be light."

Nor does the good work of Mrs. Gladstone and her daughters and daughters-in-law end with their interest in the two Homes. Hawarden Parish is large, and there is a great deal of work to be done in it. And wherever a part of this work can be performed by ladies, the " Castle ladies " are sure to undertake their share in it. They hold meetings of all kinds for mothers and daughters; they read aloud while their companions work for the poor; they visit the sick; they have poured into their sympathetic ears the troubles of the poor and lonely; they help, and cheer, and encourage wherever they can, and they do so without ostentation, as a matter of course. Thus there radiates from Hawarden Castle a great light, which shines far over the country, but which also sends some of its warmest, kindliest rays round the immediate neighbourhood; and there is not a member of the Gladstone family who does not his or her part in keeping this light uniformly bright. It is partly due to these home influences that Mr. Gladstone, in the evening of his days, lives in such an atmosphere of happy peace.

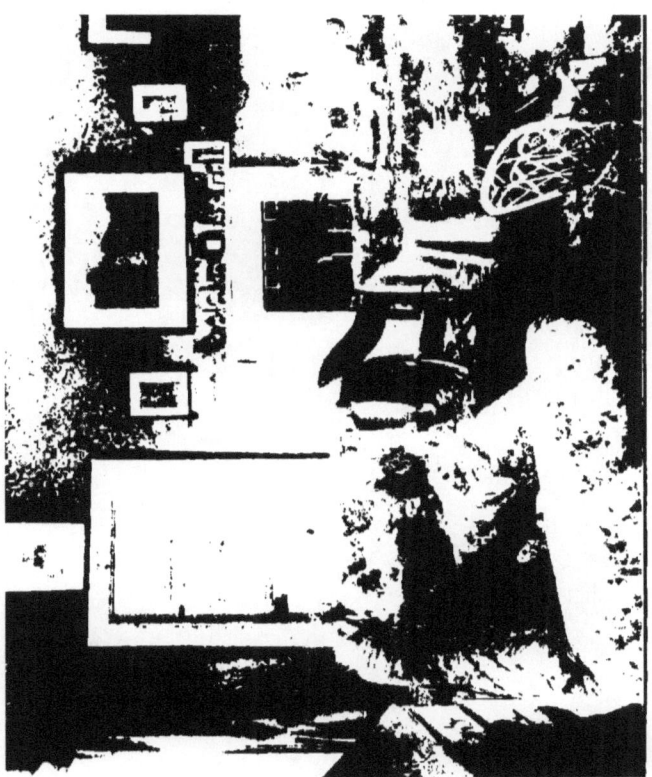

DOROTHY DREW'S DOMAIN: THE DAY NURSERY AT HAWARDEN CASTLE.

CHAPTER VI.

THE HOLIDAYS OF AN OCTOGENARIAN.

A Home from Home—Mr. Gladstone as a China Hunter—His Retirement from Public Life—The Happiest Period of His Life—An Accident in the Library—A Day with Mr. Gladstone on the Riviera—His Eyesight—His Afternoon Nap and Afternoon Tea—A Little Game of Backgammon.

Mr. GLADSTONE is one of the few happy mortals who take their home with them wherever they go, leaving only their cares and worries in the place whence they depart. Whether temporarily quartered at the historic "10, Downing Street," in Carlton House Terrace, at the Lion Mansions at Brighton, within Mr. Armitstead's hospitable walls in the north, or at a seaside hotel at Biarritz, he is always at home, and never ruffles his own or other people's temper by petty complaints concerning the absence of trivial things which it is impossible to export from a permanent to a temporary dwelling-place. It has always been like this, and many years ago, while on their journeyings abroad, his friends were anxious on Mr. Gladstone's behalf when their quarters—as hotel quarters are apt to do even in the best regulated Italian cities—left much to be desired, Mr. Gladstone himself never noticed these things. Unless, indeed, with the graceful tact which has always distinguished him, and with his extreme love of peace, he pretended not to notice them. The beds might be very far from perfection, the rooms bare and otherwise *à l'Italienne*, the food imperfect, and the service negligent, but Mr. Gladstone steered quietly, peacefully, among all these obstacles and ever made the best of things. Even inclement weather could not upset his equanimity. For, were there not museums, galleries, with more works of art than you could ever manage to study, even if you spent all your life among them?

Old china was for many years a great craze with Mr. Gladstone—one or two cabinetfuls of choice specimens at Hawarden tell of his collector's propensities even now—and he was quite happy to spend days and weeks in a place where he could " live among china." There is only one thing which has ever been known to rouse his ire when he was thus happily and pleasantly toiling and moiling among the things dear to his heart and mind. Once, now over twenty years ago, he was " deep in china " somewhere abroad, and politics, though he had left a stormy sea of them quite lately behind, had for the moment entirely disappeared from his mind. For all that those around him could tell he might never have had any closer acquaintance with the British ship of State than the doorkeeper at the foreign museum which engrossed his attention just then. But after a week among Dresden and Sèvres and Dutch and Flemish ware a political colleague, bent, like the Gladstone party, on holidaying, appeared upon the scene. He received a most cordial welcome, and basked in the smiles of his leader—we suppress, in mercy, the name of the now eminent politician who was the "hero" in this episode—till, after a while, he turned the conversation to home politics. Then the eagle eye, that had hitherto beamed so benevolently upon the new arrival, flashed fire, and the words burst forth impetuously: "For Heaven's sake, leave politics alone!" It was said in a way which made further admonitions quite unnecessary.

It is Mr. and Mrs. Gladstone's habit to spend part of the winter abroad, in the South of France, or on the sunny slopes of the Pyrenees, where, in December, 1895, he arrived on his eighty-sixth birthday at Biarritz. Mr. Gladstone enjoys these holidays immensely, but they are not holidays in the sense that they are weeks and months of inactivity. On the contrary, every moment of the day is as busy as it is at home, for the Octogenarian takes with him a great deal of work, chiefly literary, and is most careful to carry as complete a set of "tools" with him as is possible.

Few and far between are the men who have lived up to their principles concerning the division of life as entirely as Mr. Gladstone. It has always been one of his principles that an old man should pass the last years of his life in the pursuit of something more solemn than party politics. The idea jarred on him years ago that, on the threshold of another life, a man should still spend his days in political strife; and though he was by no means blind to the fine points of the man who died in harness, who fought till he fell, he had long ago made another plan for his own old age.

Thus, when the opportunity arose, and he considered himself justified in retiring from the fight, he puts his life-long principle into practice with the same firmness with which, years ago, he practised that other great principle of his, according to which no man has the right to die rich. It will be remembered that at that time he divided his possessions among his children, giving every acre of land to his eldest son, and to his other sons and daughters each his or her portion. Since then Mr. Gladstone looks upon himself not as the master of Hawarden, but as a lodger who is permitted to spend the rest of his life among his old surroundings.

When, therefore, two years ago, he suddenly retired from public into the strictest private life, and when everybody was mildly pitying him because "he must find things so deadly dull," Mr. Gladstone, as a matter of fact, was entering upon what has now proved to be one of the happiest periods of his life. He had countless interests to which he had never yet had sufficient

THE MORNING ROOM, HAWARDEN CASTLE.

time to devote himself; and without delay, without looking back, he plunged into them, as happy as a child, and became completely absorbed in his literary, classical, and theological pursuits. At that time the operation for cataract was to be performed, and Mr. Gladstone was obliged to be very careful in using his eyes. Also, an attack of influenza weakened him somewhat, especially as it left behind a slight internal weakness, necessitating a certain amount of lying up every day, up to the time of the journey to the South of France, from which, however, he returned, completely cured even of this slight indisposition. The operation on the eye, as everybody knows, was very successfully performed, but at Christmas he had an accident at Hawarden which filled those about him at first with the gravest anxiety. Authentic details regarding this fall have not reached the public hitherto, but the present account will show both the gravity of it, and also Mr. Gladstone's truly marvellous power of recuperation. He was sitting alone in his library one evening, with no one near him in that part of the house where the "Temple of Peace" is situated. He had been at his desk reading, and rose, one hand full of books, and carrying a candle in the other hand, in order to look for some book he required from another part of the room. Getting up, he forgot that he had left the lowest drawer of his writing-table open, stumbled over it, and fell headlong on the floor. Both hands being occupied, he could not protect himself, and fell with his forehead on the parquet floor. Instead of calling for assistance he tried to get up, and it was only quite by chance that one of his daughters entered the room just then. He told her what had happened, but made light of it. Both his shins were cut, his forehead had received a terrible blow, and he was badly shaken. At dinner, an hour later, he discussed Hegel in a most animated manner, then went to bed, and slept "like a top." Next morning, however, on getting up, he complained that he could not see properly, and during service at church, instead of following the Lessons in his Bible, he sat still with closed eyes. On coming home he was advised professionally to go to bed; this he did, in the middle of the day, and next morning it was found that he was perfectly restored. His shins were healing like a healthy child's, his sight was as good as usual, and the effects of the shock had passed off completely.

The following account of Mr. Gladstone's latest trip to Cap Martin, from which he returned early in the spring of 1895, is typical of all the holidays he has recently taken. It was Mr. Gladstone's own grimly humorous observation at the time that he was "very much indeed on the shelf."

The holiday in the South of France, it was hoped, would remove the last little weakness which, ten months ago, the influenza had left behind. It had obliged Mr. Gladstone to breakfast in bed at half-past eight. But staying in bed to that "late" hour did not mean sleeping. For even before the breakfast tray arrived Mr. Gladstone was absorbed in a book, and very soon after breakfast he was up and about, to read or write till luncheon-time. This was also the period when he was obliged to leave off attending Morning Prayers, together with Mrs. Gladstone, at the village church, about a mile from Hawarden Castle. To the distress of their friends, neither Mr. nor Mrs. Gladstone would ever breakfast before going to the early service, and even if a cup of tea were unexpectedly sprung upon them by anxious friends it was but very rarely that Mr. Gladstone could be persuaded to take it, while Mrs. Gladstone never broke her rule of taking nothing before going to church. Very occasionally Mr. Gladstone, before entering church, went into the Rectory to ask for a cup of coffee from the eight o'clock breakfast table.

Notwithstanding this activity, Mr. Gladstone was obliged, up to the time of his going abroad, to conform in some slight degree to the doctor's request that he should be very careful. He had to lie down after luncheon for an hour or two, and retain a reclining position as much as possible. But with his arrival in the South of France a new *régime* began. The convalescence was over; every moment of the day—including that devoted to a midday nap—was most carefully mapped out. Mr. Gladstone got up with a perfectly clear idea as to what he meant to do that day between nine o'clock breakfast and the time when, after an absorbingly interesting game at backgammon, he read Family Prayers, by the light of one candle, at ten p.m.

The breakfast hour at Cap Martin, while Mr. and

Mrs. Gladstone were staying there as Mr. Armitstead's
guests, was nine a.m. The most punctual member of the
circle, which also included the Rev. and Mrs. Harry
Drew, was Mr. Gladstone. There is never anybody
more punctual than he, as he is also one of the few
men who can dress very quickly. It takes him just
five minutes in the evening to dress for dinner, and
"dressing for dinner" does not mean slipping on a
velvet coat, or making some little fashionable change
of a similar kind; but it means with Mr. Gladstone,
donning the regulation evening dress. When alluding
to this habit of dressing quickly, Mr. Gladstone often
playfully quotes Sidney Herbert: "I take five minutes
if I potter;" because, if necessary, he can be ready in
three minutes.

Breakfast over, Mr. Gladstone, at Cap Martin, added
one new item to his ordinary occupation. He never
was a great reader of newspapers. He has described
himself as *Parcus Newspaperorum altor, et infrequens.*
But lately he has made an exception. As regularly as
breakfast-time came round *The Westminster Gazette*
arrived at Cap Martin, and the first thing Mr. Gladstone
did every morning after breakfast was to read *The
Westminster Gazette* through from beginning to end.
Next he went to his beloved books or writing table,
and with them he spent his morning in happy peace.
Mr. Gladstone, when reading or writing, wears glasses,
but he can read the very smallest type with the eye

on which the operation for cataract was performed. Thus, without glasses, Mr. Gladstone is becoming short-sighted, though occasionally, like most short-sighted people, he seems to see more than is expected of him.

One day he was walking in the Empress's Garden at Cap Martin. A gardener working among the shrubbery at some distance, and not recognising the promenaders at once, called out, drawing their attention to the fact that the gardens were private. Presently, however, he recognised the visitors, who were privileged "intruders," and took off his cap. Mr. Gladstone, though still at some distance from the man, said: "Oh, it is all right now! He takes off his cap; he's recognised us."

One o'clock was the luncheon hour, and up to that time Mr. Gladstone could but very rarely be persuaded to leave his chair. On getting up in the morning, he had mapped out every minute of the day; he knew exactly what he meant to accomplish; not a minute was wasted, and during the morning he meant to do a certain amount of reading or writing. Hence, though the sun was as golden, and the air as blue, as they can be in the South, the student could not be lured away. But once or twice, employing many a little *ruse* inspired by affection, his friends persuaded Mr. Gladstone to come for a long drive early in the day. Thus, one glorious March day, they drove thirteen miles up to where Castillone crowns one of the loveliest districts in the world. Instead of the ordinary luncheon a frugal picnic fare was stowed away in the carriage, and eaten *al fresco;* and no one was more thoroughly delighted than he whom it had been so difficult to entice away from his books, but who now entered into the picnic spirit as much as anyone.

But these occasions are rare exceptions. On the ordinary day, after luncheon, Mr. Gladstone retires for a nap into an armchair. He never lies down during the day, except by the doctor's orders. Unlike most veterans who, when found napping, protest with a great deal of (wasted) energy that they were not asleep at all, but had just shut their eyes for a moment, he readily confesses to his naps. Why should he not, since "even Homer nods?" In fact, Mr. Gladstone, in fun or seriously, heaves many a little sigh over the necessity of the nap. He is so old, he says, so weak, so very much "on the shelf," that he wants his naps during the day. And then he closes his eyes and exercises his marvellous power of calling sleep to him at will. Not for long, however, for at half-past two he is ready either for a drive or a walk. Long walks are of the past, but at the age of eighty-six he is not afraid of a tramp of from three to four miles.

Next comes the psychological moment, tea-time at half-past four. Mr. Gladstone is still as fond of tea as ever, and after having taken the air for a couple of hours it is just the best possible stimulant before another very little nap is quietly introduced. More books, more writing from six p.m. to the dinner hour at eight, or rather to the five minutes to eight, which, "if I potter," are required for dressing. Mr. Gladstone has never been a great conversationalist. He has no give-and-take, and does not adapt his talk to his hearers. But he soliloquises, like Hamlet, most delightfully. As a talker, Mr. Gladstone has always been famous, and if his "table talks" had been reported they would have made some of the most brilliant literary contributions of the century. It is well known that when a Ministerial or Parliamentary dinner was given in Downing Street, not only when Mr. Gladstone was in the prime of life, but during quite recent years, Mrs. Gladstone and her daughters, the only ladies who had been present at these political entertainments, often yearned to go back into the dining-room, whence, after they had retired, they still heard peal after peal of merriment. There is not a subject under the sun on which Mr. Gladstone cannot, if so it please him, talk with exceptional fascination.

Nevertheless, he is not always in the mood for talk. There have always been times when he was silent and absorbed, and when not even his best friends could arouse him from his thoughtfulness. This is so even now. Mr. Gladstone is not one whit less brilliant whenever he does talk; but he has periods of abstraction when "silence is golden."

When Mr. Gladstone and his friend Mr. Armitstead, or Lord Rendel, are together, the after-dinner arrangement is always a game of backgammon. It is keenly enjoyed by both players, till, at ten p.m., it is the hour of Family Prayers. At the Cap Martin Hotel, visitors who had come to see Mr. and Mrs. Gladstone and their friends sometimes asked permission to stay for Prayers; the servants, too, came in, bringing some of their fellow-domestics. Thus the circle gradually increased till after a while a considerable congregation assembled every night in the drawing-room to listen to Mr. Gladstone reading the Lessons, and to the Rev. H. Drew, who read the Prayers. After Prayers the household retired to rest.

CHAPTER VII.

WITH MR. GLADSTONE AT SEA.

Books and Sea-breezes—A Visit from the King of Denmark—Mr. Gladstone as the "Right Reverend" Gentleman—The Gothenburg System on the Spot—A Little Political Complication.

WHEN the Baltic Canal was opened, in June 1895, Mr. and Mrs. Gladstone, together with several other members of their family, were among the guests whom Sir Donald Currie took for a cruise, to see the fun. There, again, Mr. Gladstone made a most successful holiday, though when he took his constitutional on board every afternoon, at five o'clock, some of his fellow-passengers came to the conclusion that, after all, the veteran was getting old and feeble. They judged him by his slow walk and very quiet manner. And as his fellow-passengers had the good taste never to intrude upon the chief guest on board, or make any attempt at mobbing him, they had no opportunity of judging him by any other standard. For Mr. Gladstone himself, absorbed as he was in other things, was not sociable in the ordinary acceptation of the term. At this no one was surprised, since at his age any man may well be excused from entering into the petty affairs of social intercourse, even with those he knows well, to say nothing of those whom he knows not at all except by name.

There was plenty of life and enjoyment in every quarter of Sir Donald Currie's magnificent steamer, but day after day, with the sounds of genial holiday life all about him, Mr. Gladstone, as soon as breakfast was over, went into the deck-cabin reserved for him, and read or wrote all day long, with the meal-times and the half-hour's constitutional as his only intervals. The door of the cabin stood open so that he could look out upon the water, and sit, as it were, in the open air. But whenever any of his friends, in passing by, threw in a glance, they found Mr. Gladstone absorbed in the work before him. He took a few books on board; one on the North Sea and the Baltic, another concerning the Puritans, and a couple of novels for lighter reading. But these were only resorted to for recreative purposes the chief occupation being the edition of the "Works of Bishop Butler" on which Mr. Gladstone has been engaged for many years in intervals of other work, and which has now, for some months past, been the subject most prominently in his mind and most constantly in his hands.

It was in order to make some final researches at the British Museum in connexion with the life of Bishop Butler that Mr. and Mrs. Gladstone, after the cruise, stayed for a few days in London. Every evening was engaged beforehand for a dinner-party, but Mr. Gladstone spent nearly the whole of the day in quiet work at the British Museum.

It will be remembered that, just before going for the holiday cruise to Kiel, Mr. Gladstone was indisposed. At the time no one outside his own home circle and his physicians knew how grave this indisposition was. And it was as sudden as it was serious. One evening he went to bed perfectly well, and next morning he woke up with his voice entirely gone. Later on a terrible cough and great feverishness heightened the anxiety of Mr Gladstone's friends, and even the day before starting for London, and for the *Tantallon Castle*, it seemed very doubtful whether the voyage could be undertaken at all. However, a sudden change for the better set in, and Mr. and Mrs. Gladstone, together with the Rev. and Mrs. Harry Drew, left Hawarden Castle at the appointed time. Mr. Gladstone, who is an exemplary patient, was perfectly free from anxiety about his state of health, but one thing troubled him not a little. Like the fall he had last winter, the cough

now had affected the sight of the eye on which he had
recently had the operation for cataract, and objects
before him seemed indistinct and confused. Considering how much Mr. Gladstone depends for his keenest
interests and pleasures on being able to read and write,
this weakness of the eyes was very troublesome.
Fortunately, soon after the beginning of the voyage,
it disappeared entirely, and Mr. Gladstone saw again
quite as well as before.

Many and pleasant were the impressions which
he brought back from his cruise to the North. One
of the things which touched him very deeply was
when, on meeting the King of Denmark, the latter was
so moved that his voice broke, and he could hardly
complete his speech. The banquet of enthusiastic
Hamburgers on board was greatly enjoyed, and the
Burgomaster's amusing little mistake in addressing
Mr. Gladstone as the "right reverend"—a mistake, by
the way, which was altogether unintentional—was
appreciated at its full value by the "right reverend"
gentleman.

No sight-seeing of any kind was done by Mr. Gladstone
and his friends. At Copenhagen there was just a drive,
on a very cold afternoon, to a beautiful deer park
some distance from the town. But Gothenburg, that
most interesting of ancient Norse cities, aroused
Mr. Gladstone's deepest interest. He drove about,
and even went so far as to study the Gothenburg

system on the spot by entering one of the public-houses, and informing himself of the details of the liquor trade as they appear when actively applied. An amusing little incident took place on the same occasion. One of the most interesting places in Gothenburg is an old curiosity shop of great importance. To this Mr. Gladstone went with a few of his friends, only to find, on arriving, that the owner was "away all through June," as a label informed all would-be purchasers. Poor shopman! The news which reached him on his return, of the visit he had missed by going "away all through June," would no doubt spoil his recollections of the summer holiday of '95.

From the mysterious source whence flow at all times, if a personage be prominently "before the public," the most astonishing tales and stories, there came, at that time, one that attributed to Mr. Gladstone the saying that it was for the sake of his health, and not for his enjoyment, that he had accepted Sir Donald Currie's invitation for the cruise to Kiel. Seeing that all through his long life Mr. Gladstone has been the very pattern of tact, politeness, and stately courtesy, it is almost incredible that so clumsy and ungracious a statement could be attributed to him. As a matter of fact, Mr. Gladstone was most grateful to Sir Donald for the very enjoyable holiday, and took endless trouble to compose a note of thanks to Sir Donald in his name and in the names of the other guests on board. The exact wording took him some time, and when he was satisfied with it he sat down in his deck-cabin and copied the whole of the letter, so that it might be given to Sir Donald before the party dispersed. At breakfast, on the last morning, he read the letter out, preceding it by such graceful words of appreciation of the hospitality on board the *Tantallon Castle* that this little impromptu speech is reckoned by not a few who heard it as the best of Mr. Gladstone's speeches during the cruise.

Another unreported speech Mr. Gladstone made while on board was to the Chief Engineer of H.M.S. *Royal Sovereign*, who, together with his fellow officers, came on board to be presented to Mr. Gladstone. The subject of this eloquent speech was an enlargement on the wonderful work, order, and organisation which made up the discipline on board a naval vessel.

The only way in which the political crisis which followed Mr. Gladstone's return immediately affected him and his friends was rather amusing. It had been arranged that on their return to town Mrs. Drew should stay at the house of Mr. and Miss Balfour, who are among the most intimate friends, in private life, of Mr. Gladstone and his family. On reaching Gravesend, and hearing the news of the crisis, Mrs. Drew decided that, under the circumstances, it would be better for the comfort of everyone concerned that Mr. Gladstone's daughter should not be a guest at the house of friends in the very front ranks of the enemy's camp, and thus in the very thick of the crisis, with its incessant calls and missives. A wire was, therefore, sent Miss Balfour, to the effect that Mrs. Drew was going to the house of other friends. A few hours after her arrival there Miss Balfour called, and it was agreed that just then, notwithstanding the most cordial relations, it was a relief that an invitation given in times of peaceful calm could be cancelled on the day of storm and stress.

CHAPTER VIII.

MR. GLADSTONE'S GIFT TO THE NATION.

Mr. Gladstone on "Posthumous Charity"—The Sale of His Collection—His Determination to Found a Theological Library—The Scheme for St. Deiniol's Hostel and Library—Growth of the Scheme—Mr. Gladstone as Librarian—A First Glance at the Library.

THE first and foremost subject which occupies Mr. Gladstone, and of which he never grows tired, is, as has been said, his theological work. Bishop Butler comes first, and other theological studies are almost as absorbing. But besides these, there is one subject on which Mr. Gladstone has spent a great deal of thought and time and care since he has retired from public life. This subject is St. Deiniol's Library, Mr. Gladstone's magnificent gift to the nation, which passed formally into the hands of Trustees on January 1st, 1896.

It is not known to anyone when the idea of the Hawarden Theological and General Library first entered Mr. Gladstone's mind. But the making over to some institution for the use of the nation of his rich store of books has long been decided upon. Also, that the gift should be made in Mr. Gladstone's lifetime, for, like that other Grand Old Man, Mr. G. F. Watts, R.A., he does not think much of what he calls "posthumous charity." It is easy enough, Mr. Gladstone reasons, to give away your possessions after you are dead, when you want them no more. But the right thing is to do so while you are alive; to give to others as soon as you can do so, instead of keeping it all in order that you may increase your own comforts and those of your family.*

---

* In this connexion it may be interesting to recall a letter which Mr. Gladstone wrote shortly after the publication of Mr. Andrew Carnegie's "Gospel of Wealth." We sent the proof-sheets of the article in which Mr. Carnegie discussed, in detail, his doctrine that "the man who dies rich dies disgraced," to Mr. Gladstone, and the following was the reply:—

HAWARDEN CASTLE,
Christmas Day, 1889.

DEAR SIR,

I answer your letter promptly, for I think it an honour to be associated with Mr. Carnegie in the noble purposes and doctrines of the " Gospel of Wealth." I have, therefore, begun by reading his proof-sheets, in violation of a rule I am obliged frequently and ruthlessly to apply to other quarters.

More than twenty years ago, when Mr. Gladstone sold his fine collections of china, old ivory, etc., he began to put this theory into practice. A tiny remainder of these treasures may still be seen at

---

1 follow Mr. Carnegie in nearly everything he affirms and recommends. My main reservation is prompted by his language respecting endowment of twenty million (dollars) granted with a splendid munificence to Stanford University.

My mind is possessed with much misgiving, which rather grows than diminishes as years roll on, about the wholesale endowment of offices and places. That subject is much too large for present discussion. I may say, however, that there is a large field for University expenditure lying beyond the scope of this remark; the remark itself I may illustrate by saying with reference to the large endowments of learning, or, at least, of places and offices with a view to learning, in this country, that I have doubts whether it does not raise the market-price of the higher education, which it aims at lowering, and which I suppose to be dearer in this country than perhaps in any another. I do not, however, go beyond this, that I cannot join in an affirmation quite so unqualified, on this particular point, as that of Mr. Carnegie's article.

I must add that the growing tendency to the dissociation of Universities, as such, from religion does not abate, but enhances, the force of all such considerations as have suggested my language of reserve.

I now come to an important addition which I should like to attach to the Gospel of Wealth. I see no reason why, in the list of admissible and desirable objects for the dedicatin of friends, we are not to include their direct dedication to the service and honour of God. The money spent in the erection of our cathedrals, and our great churches hardly inferior to cathedrals, has been large, and has in my judgment been very well laid out. What I have said as to the endowment of

Hawarden Castle, in a few cases containing specimens of precious porcelain, carved ivory, and exquisitely beautiful old Italian jewellery. A little incident which happened the other day shews that Mr. Gladstone did not find it an easy thing to part with the things of beauty acquired in the course of many years. Mr. Herbert Gladstone brought home a carved ivory figure which he had chanced to see in an antiquarian's shop at Liverpool, and which was labelled "From Mr. Gladstone's Collection." Mr. Gladstone recognised it at once, and presently the price was mentioned which his son had paid for it. The question then arose whether the figure had risen or fallen in value, and somebody suggested that Mr. Gladstone had a catalogue sent to him at the time of the sale of his collection, in which the prices were given. "Yes," he replied, " the catalogue was sent, and I have it still; but I have never

---

offices and places has some application to the great province of religion. But, apart from this, and apart from marvellous and noble works, such as the cathedrals, the institutions of religion, the works of devotion, learning, mercy, and utility, connected with it, are numerous and diversified. Religion is a giant with a hundred hands, whose strength, however, is not for rapine, but for use. I should wish to bring its claim, proportionate and therefore large, under the consideration of the open-handed and open-minded philanthropist.

    I remain,
     Your faithful and obedient,
       W. E. GLADSTONE.

had the heart to look at it." This was the only time that he had ever said one word about his private sentiments with regard to the sale.

As the years went by the idea gradually took shape of making over to the nation his own library. When this decision had been taken, Mr. Gladstone began to look about for a fitting place where the books might find a home. At first it was thought that rooms might be set apart for them in an existing institution, perchance in London or Liverpool. This, of course, would have obviated the necessity of acquiring a special building. But it was argued that neither London nor Liverpool, nor any other great central town, would be a place wherein quiet students and scholars might with advantage pursue a course of study in "Divine learning." The restlessness and roar of millionfold human life would be a disturbing element in any library intended for a Temple of Peace. Mr. Gladstone fully agreed with this objection, and it was then suggested that, since he now entertained the idea of causing a special building to be erected in some rural district, Hawarden itself would be the most suitable spot. The perfect seclusion of the village; the ease with which it may be reached (especially now that a new line has been opened between Liverpool and North Wales, which brings Hawarden within half an hour's distance from Liverpool); the beauty and healthiness of the district; and also the associations

H

of the whole place with Mr. Gladstone; all seemed to point to Hawarden as the ideal situation for the Theological Library.

Some four or five years ago Hawarden village awoke one morning to an interesting fact. A new building was to be added to the place, and such a fact, in villages much larger than Hawarden, is always important and exciting. Not far from the famous old Parish Church, nearer still to the ancient building known as the Grammar School, the stonemasons had begun their work of laying a brick foundation. Who was the enterprising spirit that had chosen this spot at the end of the village, on the edge of the hill, for a dwelling-place?

At first there were manifold rumours, but gradually it appeared that this was not going to be an ordinary house of brick or stone, as is every self-respecting cottage at Hawarden, but an iron structure. What! a chapel-of-ease, so near the old Parish Church? Or—but surely this could not be so!—a Dissenter's chapel? It was neither; it was a library, and it was Mr. Gladstone's. And with this knowledge the villagers had to be satisfied. But when the Iron Library was finished, when cartload after cartload of books had been sent to it from "The Castle," when builders and carpenters and painters had finished their work, one more item of information was added to the villagers' knowledge on the subject. The library was not for them. This might have been a blow—even though

Hawarden had its library at the local institute—to some spirits thirsting for "book-learning;" but the blow was tempered when it became known that the bulk of the twenty thousand volumes of which it is composed are mostly on theological and kindred subjects.

ST. DEINIOL'S LIBRARY.

"The Castle people," it was evident to all Hawarden, took the greatest interest in the Iron Library, and Mr. Gladstone himself, whenever he fled to Hawarden Castle in his short intervals of rest, spent a great many hours among the books, which he himself had selected from the treasures of his own library. In the library at the Castle, meanwhile, there were also abundant signs that something special was going on. The "Temple of Peace," as Mr. Gladstone's library is always called by his family and friends, looked as if some great upheaval were taking place. The floor and the tables were covered with boxes and parcels of books, all "weeded out" from among the rest by the owner himself, and all destined to fill the ample shelves of the new Iron Library.

Thus gradually the project of the St. Deiniol's Library took definite shape, and together with it grew up the twin project of the Hostel, which also goes by the name of the same old Welsh patron saint. For if this retreat for theological scholars and students were situated far from the madding crowd, in a quiet Flintshire village, it must needs be joined to some place where the students could make a congenial temporary home. Hence the Hostel. When the spot for the present Library had been fixed upon, at the end of Hawarden Village, the house which was to be converted into the Hostel was already there. It had been a grammar school for many a long year, but now it was empty, and as yet no intending tenant had made a bid for the quaint old brick house, with its solid oak staircases, its desks and peep-windows and raftered ceilings in the large old class-rooms. Mr. Gladstone's daughter, Mrs. Drew, who, together with her husband, the Rev. Harry Drew, had taken the liveliest interest in the whole scheme from the very beginning, saw at once that the whilom grammar school (now a thing of the past since the Intermediate Education Act destroyed all institutions of this kind) would prove a veritable treasure, an ideal hostel. And there and then Mrs. Drew set to work, and converted the dilapidated old place into as delightful and as

VIEW OVER THE DEE FROM ST. DEINIOL'S LIBRARY.

picturesque a dwelling-place as can be imagined. Even eighteen months ago, just after the first students had availed themselves of a quiet "theological holiday," the place was full of charm, standing in a green country garden full of flowers and shrubs and old fruit trees But now, after it has been in full working order for more than a year, and after over a hundred students have lodged within its walls, the Hostel blossoms literally like a rose. Everything is simple. If it were not, it would be out of harmony with the whole aim and object of Mr. Gladstone's gift. There is a little landing upstairs, with just a few chairs and a table, and with a magnificent view over the low-lying district over which, after dark, the lights of Liverpool give a lurid colouring to the sky, where Mrs. Gladstone often comes to write some of her letters. "It is so peaceful up there," she says, in her kindly way; "I can get on better than in almost any other place, when I sit by the window with that view before my eyes."

For a student coming from the busy world, and with his mind filled with the grave, great subjects with which are filled the majority of the twenty thousand volumes in the adjoining Library, the place is nothing short of the ideal. And as the interior of the Hostel, so its surroundings. The jessamine and ivy which hung in wild liberty over the front have now been trimmed and clipped into perfect neatness, and send their shoots up to the very roof. The garden, with its hips and haws on all the rose-bushes, and with innumerable signs (understandable to every lover of a garden) of last summer's blossomings and bloomings, is beautiful even in December, and in its sheltered nooks the velvety pansies have not grown tired of keeping, with grandmotherly faces, a watch over the place, and giving a welcome to all who may walk along the gravel walk that leads from St. Deiniol's Church to the Hostel, and then onward to the Library.

As it is with the Hostel, so it is with the Library itself. In one short twelvemonth it passed out of its initial stage into that of perfect order and regularity. There was at the beginning only a somewhat undefined country path across the green field separating the Library from the high road. Now a wide, neat, gravelled walk, at the end of which a new iron gate prevents the "man in the street" from too easy admittance into the precincts of the Library, leads up to the very door. The roses in the beds under the Library window are taking root and courage, and in the lawn the crocuses of next March are asleep. Up the sides of the iron building the climbing plants are lustily growing, and what was a rough field is now an even lawn.

It is true the present iron building in which the Theological Library is housed is to be only a temporary home, but those in authority have wisely avoided the great mistake so often made, where "things are only

A WOODLAND HOLLOW IN THE PARK.

temporary," of altogether neglecting appearances and comfort. The iron building might be intended to last for centuries, so much care and attention have been bestowed upon it in every way.

The two chief rooms in the Library have now a settled look about them. They are two rooms as beautiful as a great many good books, well printed, well bound, and well arranged, can make any room. You see at St. Deiniol's Library the truth of the saying that nothing furnishes a room so well as books. Not that there is any lack of other suitable furniture to give a cosy and homelike look to the rooms. There are good large substantial writing-tables with all the necessaries for note-taking, etc.; there are easy chairs in the recesses between the windows, and there are some interesting pictures, medallion portraits, etc. One of the rooms has been named by Mr. Gladstone the Divinity Room, the other the Humanity Room. The former contains only works on theology and kindred subjects; in the other are included books on most subjects under the sun, many of them being intended for recreative reading, such as Mr. Gladstone himself indulges in after he has "worked at books" for ten hours a day.

Dr. Döllinger's portrait hangs over one of the doors, and a small terra-cotta medallion of Cardinal Manning was hung up by Mr. Gladstone against one of the shelves on which the works on Roman Catholicism have their places. Over the door leading into the Humanity Room a fine cartoon by Henry Holiday has recently been placed. The female figure aptly represents " Theology."

But it is now time to look a little more closely into this remarkable Institution, and to explain its objects. This, with the aid of facilities most kindly afforded us at headquarters, we proceed to do in the following chapters.

CHAPTER IX.

ST. DEINIOL'S LIBRARY.

The Inception of the Library—The Object of the Foundation—A Dream of the Future—The View from St. Deiniol's—A Student's Retreat—The Divinity Room –The Humanity Room—The Warden's Study—The Trust Deed—Mr. Gladstone at the Library—Mr. John Morley and the Photographer—Mr. Gladstone's Eyesight.

A FEW years ago the *embarras de richesses* in books rendered it necessary that something should be done to prevent overcrowding and its attendant annoyances. It was then that Mr. Gladstone disclosed his scheme of making over to the nation the Hawarden Library, which, of all his possessions, he values most. The great mass of important theological works (works of which each volume has a deep interest to Mr. Gladstone himself, great theologian as he is) would be all but wasted in an ordinary private library. Hence, they must not be allowed to remain private property, but must be disposed of on the principle of making them of the greatest use to the greatest number.

Mr. Gladstone's scheme for securing this end is as follows:—In the days to come, he argues, should the Disestablishment of the Church of Wales become a fact—when an endowed clergy would cease to exist—it would be a boon to theologians, students, scientists, and those connected with any educational work, to know that there is a place to which they may go for those works which they themselves are unable to procure. Considering who is the originator and founder of the scheme, it goes without saying that the whole scheme is on the broadest and most liberal lines, and that the Library is founded for all students and readers, absolutely regardless of denomination. That this is a generous scheme will strike one and all, but how generous it is can only be realised by those who have an idea of the truly magnificent Theological Library, including many of the most rare and valuable works, which Mr. Gladstone thus puts at the service of the the public.

The Library itself, however, is only one part of Mr. Gladstone's scheme. Those who go to a theological library for instruction or research, require, as a rule, a course of study extending at least over a few days, and in many cases over weeks, and perhaps months. Hence they require hotel accommodation, or an equivalent for it. But hotel accommodation even, under the most favourable circumstances, and apart from pecuniary considerations, is not exactly in harmony with the requirements of a student of theology or kindred subjects. He wants quiet for thought and reflection, even during the hours he is not actually at the Library; he wants, if possible, kindred spirits, with whom, at meal-times or in a leisure hour, he can discuss subjects of common interest. And he wants also, in order that he may preserve his bodily health, a place where his material wants may be cared for in a simple, sensible way.

This want, also, has been taken into consideration and fully supplied by Mr. Gladstone. St. Deiniol's Hostel, standing between Hawarden Parish Church and the Library, and less than a stone's-throw from either, has been instituted for the accommodation of those wishing to make use of the Library. We shall presently give a full and detailed account of the Hostel, but return for the present to the Library itself.

It stands high on the breezy hill-top, this Iron Library, which contains twenty thousand volumes, many of them counting among the rare literary treasures of the country. It is not a thing of beauty, seen from the outside, though its somewhat irregular form, its small spire, and the climbing plants and flowers around it, which are just beginning to take courage and grow, redeem it from the hopeless unsightliness afflicting most iron buildings. One step off the village street towards the Library, and you are in the field in which the building stands. As you look around you, from the Library door, you think that it would be almost a pity to improve away the scarlet poppies in the grass, the daisies, and the golden dandelions: the simple naturalness seems so entirely in harmony with the place and its object.

Then, turning from your immediate surroundings, you look farther afield, and you are satisfied. On your right, on the hill slope, is Hawarden Churchyard, where, for some centuries past, the Flintshire villagers have been laid in their last sleep. Of course you must presently go and look more closely at the graveyard, on your way to the church made famous by Mr. Gladstone; but from where you stand, at the library door, only one grave attracts your special attention. A simple white marble cross rises from it, under a young acacia tree; there are no other graves near; a wreath of deep blue velvet pansies circles the grave, and on the front of the gleaming marble cross you read the name. "W. H. Gladstone, who passed away 4th July, 1891," and under it, in black letters:—

> Soon shall come the great awaking,
> Soon the rending of the tomb;
> Then the scattering of all shadows,
> And the end of toil and gloom.

You turn to the left, and your eye falls on a rounded hillock, with a crown of thriving pines and other trees

This also is a grave, but one of ancient times, and no one knows to what hoary warrior may have belonged the bones that were found beneath this picturesque green mound.

The hill slopes rapidly from the Library field. The school is just below; a few belated lads straggle towards the door; from the inside comes faintly the sound of a song chirped by a chorus of shrill infant voices, and silently, close by, with a cloud of clear blue smoke rising from it, lies the schoolmaster's idyllic abode, smothered in roses and clematis, blooming amidst the blaze of scarlet and gold foliage of a climbing plant. Beyond the cluster of trees you see gabled cottage roofs. The gardener has left his ladder, from the heights of which he has been clipping the too luxurious jessamine branches covering the front wall of the Hostel. "See those houses?" he says, pointing towards the gables under the trees; "my mother lives in one of them. She remembers when Mr. and Mrs. Gladstone were married, fifty-seven years ago. She was a servant in this here house, which is now the Hostel. The two Miss Glynnes was married on the same day; my mother saw it all. And there was open house, yes, here at the Hostel, all the day and all the night. Anybody might come in and eat and drink as much as they liked. Does my mother remember it all? Bless you, same as you would. Trust her for being all there." And then he sidles a little closer up to you. "Is it true what they say, that *he*—" pointing his thumb over his shoulder in the direction of Hawarden Park gates—"is coming here this afternoon? About three? Then, ye'll see, he'll be here on the stroke of three. If he's walking, he'll come through the churchyard, and if he's driving, through the field. If he says three it will be the stroke of three. He's as good as his word."

You turn to the view once more. The autumn mist has lifted; for a moment you realise that from this breezy hill you have a magnificent view: magnificent for its extent over the green wooded plain below, "across the sands of Dee," to where, in the far distance, the pale sea joins the paler sky; magnificent also for its perfect repose and peace. Truly, an ideal spot for a student's "retreat."

A bright fire meets your eye at once as you open the Library door. Everywhere about in the large room are books—books—books. The Iron Library is arranged in the same ingenious way as Mr. Gladstone's private library at Hawarden Castle. There are windows on either side of the long room, and between these windows high bookcases, running towards the centre of the room, are put up. There are books on either side of these cases, and the part facing the centre of the room is again arranged to hold books. It is truly marvellous how many books can thus be stored without a single one being out of sight.

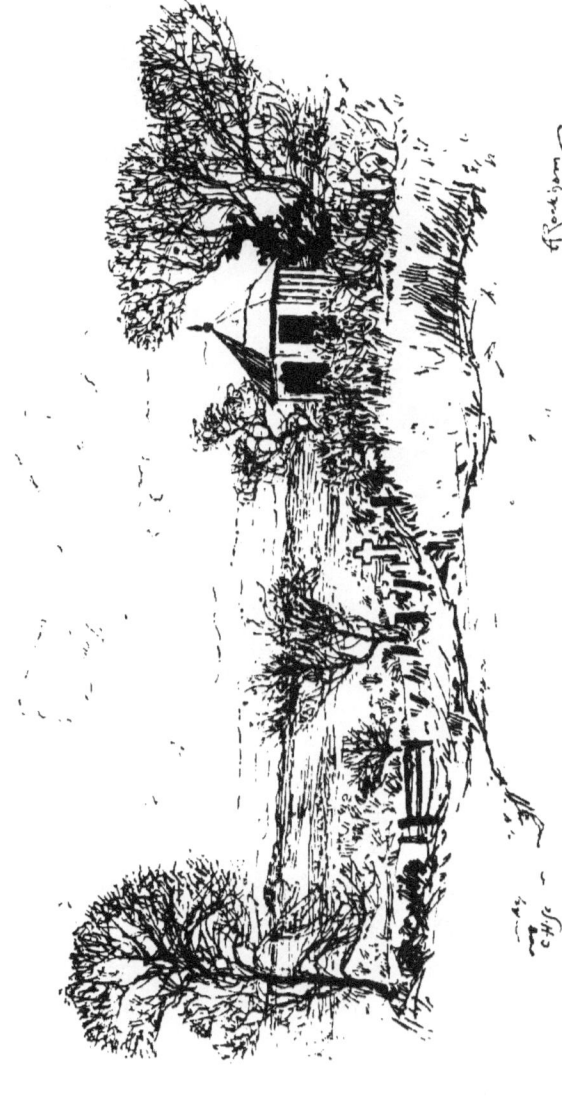

SUMMER HOUSE, AND VIEW FROM ST. DEINIOL'S LIBRARY.

Apart from this arrangement of the books, the room reminds you strangely of the "Temple of Peace" at Hawarden, where Mr. Gladstone has spent so many of the happiest days and hours of his life. There is the same simplicity, the same quiet comfort, the same air of repose, and the same absence of library conventionality about. Such is the Döllinger or Divinity Room, the most important part of the Library, and as your eye wanders over the splendid army of books, you almost envy the students who may come here to read to their hearts' content. The only rule concerning the contents of the books is, that none of Mr. Gladstone's annotations on the margins may be copied or quoted as illustrating his views on certain questions, since such quotations might convey an altogether wrong impression. As a matter of fact, Mr. Gladstone often jots down marginal remarks when an idea occurs to him while reading, though that idea may in no way represent his views.

Through a door, over which hangs the fine cartoon by Henry Holiday, you reach the second room in the Library, to which Mr. Gladstone has given the name of the Humanity Room. It is arranged on exactly the same plan as the first, and contains secular works chiefly. You note Madame de Sévigné's letters on one shelf, in neat and dainty little volumes; a yellow-backed Zola lower down, among miscellaneous French works; six gigantic volumes of "Denkmäler aus Egypten," which have just been bought from the great Quaritch for sixty pounds, are being unpacked; all your favourite English and foreign authors and poets are there: biography, history, travel, fiction—in fact, whatever your heart may wish for in the way of literary wares, is here at your disposal, in order that when you have steeped your mind in theology during a long, studious day you may have some recreative reading provided in the secular room. Many an object in the Hostel and Library is interesting for its association with Mr. Gladstone's public career. Here, there, and everywhere you see caskets, paper-knives, albums, chairs, pictures, and ever so many other things, all of which have been given to Mr. Gladstone on some occasion by political adherents and disciples. In the Library itself there is a fine walnut book-case, which, with excellent taste, the Leeds, Huddersfield, and Heavy Woollen District Federation of Liberal Clubs chose in the year of grace 1895, as a present which would probably give more pleasure than anything else to their G.O.M.

Five small private studies have been set apart in the Library for such students who may prefer to do their work in solitude. Each of these rooms is a cosy little corner with everything required for a quiet reading and writing room, and with a view from the windows which is beautiful enough to make any student lift up his heart, however puzzling and worrying his knotty points of theology may be.

ST. DEINIOL'S LIBRARY: THE DIVINITY ROOM.

Next there is the Warden's bright little sanctum. The Rev. H. Drew (who holds the combined office of Warden and Chief Librarian of St. Deiniol's), and the Assistant Librarian (Mr. Rye), have a working period behind them of which they may well be proud. Mr. Drew, after helping Mr. Gladstone with his correspondence arrives every morning shortly after eleven at the Hostel. All the twenty thousand volumes in the Library had to be catalogued, and what this means no one will fully appreciate unless he has ever tried his hand at cataloguing books in a satisfactory manner. The Divinity Room catalogue is complete, and it is a pleasure to open a drawer in the neat library card catalogue cabinet, holding the Ceres Card Files, which are far and away the simplest and most satisfactory invention of their kind. Considerable additions (which fortunately can be made to the card catalogue without at all interfering with its neatness and comprehensive form) will have to be made whenever the remainder of Mr. Gladstone's gift books, now still at the Castle, are brought over to St. Deiniol's. These number from six to eight thousand volumes, all of which will eventually go to the Theological Library. It is interesting, by the way, that the library at Hawarden Castle—is by the owner thereof now only considered as part of St. Deiniol's Library. It is simply, as we have before stated, "the Castle section," which Mr. Gladstone looks upon as a loan he is privileged to make from the rightful owners, the men to whom he has bequeathed every one of the books he has acquired in a long lifetime. The books

ST. DEINIOL'S LIBRARY: THE WARDEN'S ROOM.

of the far-famed Glynne library, which he found when Hawarden came into his possession, are the only part of his library which he does not feel at liberty to dispose of.

There is one interesting piece of furniture in the Warden's room: a large carved oak armchair, with a small round plate let in at the back, bearing the following inscription: "To the Right Honourable W. E. Gladstone, M.P., First Lord of the Treasury, by the Liberals of the Borough of Greenwich and the Liberal Clubs of the neighbourhood. In testimony of their high appreciation of the priceless services rendered by him to the country, and in remembrance of the proud distinction he conferred upon the borough as its Representative in Parliament from 1868 to 1880."

With January 1st, 1896, St. Deiniol's Library and all that is comprised in Mr. Gladstone's gift to the nation entered upon a new phase of its existence. It passed formally into the hands of the eight Trustees whom Mr. Gladstone has paid the compliment of asking them to accept this office from him. The first informal meeting of the Trustees was held in October 1895, at Hawarden Castle, Mr. Gladstone presiding, and setting forth the whole scheme. He himself had sketched the Trust Deed, which was drawn up by Sir Walter Phillimore, one of the trustees, and which since then has been somewhat altered in form, but very little in substance. The main features of the Deed are as follows:—

The property purchased by Mr. Gladstone—including not only the present Hostel and Library, but a large field immediately adjoining—is vested in the Trustees, together with all the books and papers now in the Library, or hereafter to be conveyed thither by him.

A sum of about thirty thousand pounds in various stocks is included in the settlement for maintenance and endowment purposes. The nature of the trust is thus defined :—

> To form a library and institution, to be called St. Deiniol's Library, which library and institution is to be for the promotion of Divine learning in the sense put upon those words by the Church of England, with which Church the Settlor, believing that she will be always faithful to her trust, hereby provides that the Foundation is always to be in connexion.

The chief official of the Foundation is to be called the Warden and Chief Librarian, and is to be in Priest's Orders. During the lifetime of the Founder the Warden is to be appointed by him; and "the Trustees shall endeavour, as far as possible consistently with the main design, to maintain the connexion of the foundation with the family of the Founder and with the Parish of Hawarden." Two further points in the Trust Deed may be mentioned as of some special interest for their "liberal" tendency. One is that the Trustees and officials may be men or women, married or single. The other is the ample power given to the Trustees to make the Foundation "move with the times." They may alter its form, divide it, and remove it from the Parish of Hawarden—though this latter power is not to be exercised for at least five years—or, by consent of all, dissolve it.

THE REV. H. DREW, WARDEN OF ST. DEINIOL'S.

The actual amount set apart by Mr. Gladstone for the endowment of the Library will bring in a considerable income. Added to this, there will be the income derived from such students as can afford to pay the sum of twenty-five shillings a week for board and lodgings at the Hostel. It is believed that thus there will be sufficient money to defray the expenses of the institution, to maintain a special fund for poor students and clergy, who upon inquiry are found to be unable to pay for their stay, and also to set apart a sum of money for eventually erecting a more permanent building wherein to house the Library. The surplus, after this has been accomplished, is to be devoted to the purchase of further books suitable for a collection of works on Divine learning.

It is Mr. Gladstone's wish that the permanent building should be in the same place where, in the present iron structure, the scheme came first into operation. For this purpose he has recently bought from the Ecclesiastical Commissioners the Hostel and the glebe-land upon which it stands, and from the Hawarden estate an additional large piece of ground adjoining the Hostel and surrounding the Library. The Trustees of the fund are all personal friends of "the Settlor," and fully in sympathy with his views. They are the Rev. Stephen Gladstone (Rector of Hawarden), Countess Grosvenor, the Hon. and Rev. A. Lyttelton, Sir Walter Phillimore, the Hon. Mrs. W. H. Gladstone and

Mr. Henry Gladstone, Mr. George Russell, and Mr. Charles B. Toller, of Acton Bank, Hawarden. It is provided in the Trust Deed that the Rector of Hawarden, if a Trustee, shall be Chairman of the body.

THE RECTORY, HAWARDEN.

Thus is will be seen that in this his munificent gift Mr. Gladstone has acted on the same large-minded and large-hearted principles to which he has adhered all the days of his life. May he live to see St. Deiniol's Library yield fruit a thousandfold!

Here is one little scene at the Library, which, at all events to those who watched it, gives the institution a separate and particular charm, apart from its own attractions. The grey October afternoon lies on the quiet village of Hawarden and on the hill, on which the Library, the Hostel, and the Church stand close together. The ivy-covered door in the churchyard wall opens, and Mr. John Morley, together with Mrs. Drew, walks towards the Library. Mr. Morley is on his way back to Dublin, and, as usual, has broken his journey to spend a few hours with the old chief. Presently a carriage drives into the Library field, and before any of the other occupants have had time to descend, a vigorous figure, wearing a thin autumn coat and soft felt hat, is approaching the Library with eager step. It is the G. O. M., with face as ruddy from the crisp air as an apple, and, with eagle-eyes, taking in everything around. He walks erect, and looks in better health than he has done for some years past.

Has he promised to be specially photographed for the *Westminster* account of the Library scheme? Very well, then it must be so. Stand here, at the door? Hat off or on? Look at those yellow leaves? And there he stands, hat in hand, motionless; the Octogenarian, to whom to be photographed is anything but a pleasure, but who subjects his own wishes to those of others. The keen wind touches his thin white hair, and quietly he obeys the photographer's instructions— not once, but several times—with only the request that he may not be kept standing too long. Behind him, Mrs. Gladstone Mrs. Wickham Mrs. Drew, and

Mr. Morley are waiting in the library, but there is one member of the party whose patience gives way, and who sees no reason why he should not be photographed as well as Mr. Gladstone. Mr. John Morley has put away the wily photographer's suggestion that he also should be photographed, with the smiling remark: "Not the great and the small together. Let the great be alone." But Petz, Mr. Gladstone's black Pomeranian, has no such scruples at all. His small muzzle is between the chinks of the Library door, and very energetic he is in his growled protests that he wishes to be outside beside the master whose constant and faithful companion he has been for many years. Poor old Petz! His little black head is turning white, but he is like his master in this, that though he is growing old in years his spirit is undaunted, and to his energy there are no bounds.

A moment after the photographer has released Mr. Gladstone you look about in vain for the figure of the G.O.M. The carriage is still waiting in the field; Mrs. Gladstone and her two daughters, Mr. Morley and Canon MacColl, are wandering about in the Library—but where is the Master? Look into that window niche in the Divinity Room. There, in a basket chair, with Petz at his feet, and with an expression of perfect repose and peace about him, he sits quite absorbed in a small volume he is reading. He reads without glasses, apparently without any difficulty. As a matter of fact, Mr. Gladstone's eyesight is now quite restored. Both his eyes are perfectly clear and bright, and the only outward and visible sign of the recent operation for cataract on the one eye is a thin white ring round the pupil. Since Mr. Nettleship, on his last visit to Hawarden, pronounced the cure complete, Mr. Gladstone's patience has been somewhat severely tried by the host of different glasses, which to save his eyes as much as possible, he has been recommended to wear. There are glasses for reading small type, and others for large type; glasses to look at objects in the room, and others at objects at a distance. Sometimes Mr. Gladstone forgets his eyes and spectacles altogether, as was the case when he was reading that afternoon in the Iron Library; but he is too obedient a patient not to carry out every order of his medical attendant once the reason of an edict has been explained to him.

If Mr. Gladstone were fifty years old instead of eighty-six, the other eye—which, as generally happens in case of senile cataract, is affected in the same way as was the eye which has now recovered its normal sight—would undergo the same operation.

After a while Mr. Gladstone's reading is interrupted. An interesting little group gathers round him. Mrs. Gladstone, Mrs. Wickham, and Mrs. Drew, in their unwearying efforts to save him from over-exertion of any kind are drawing his attention away from his book; Mr. John Morley leans against the bookcase, opposite the armchair, and engages his late chief in conversation, and presently the whole party, Petz well ahead, is making its way towards the door. Truly, with that flash in his eyes, that ruddy tinge on his cheeks, that dignified and at the same time graceful courtesy to strangers, Mr. Gladstone is a perfect type of a man of culture in his green old age.

Outside a small crowd has gathered. They stand, hats doffed, and bow their silent salutations to him. " I've come all the way from Ireland on the chance of seeing Mr. Gladstone," explains an enthusiastic Paddy over and over again, evidently in the hope that such devotion might possibly be rewarded with the high honour of shaking hands with the Grand Old Man. Mr. Gladstone's friends try to keep the Irishman off, but, pulling his carroty locks, he steps at last boldly up to the open carriage and gets his reward, Mr. Morley sitting by and watching with his quiet smile the little scene. And if you have heard much of the Old Man's physical infirmities, you will laugh at those rumours after you have felt his strong and hearty handshake, and agree with the beaming Irishman, who, after the carriage had left, threw his hat into the air and sang out right lustily: "More power to him, bless him! He's more life in him than most men half his age!"

When the day is declining, and before it gets too dark for reading, Mr. Gladstone often leaves his library, taking his book with him, and sits on the stone steps leading from the Castle library on to the front lawn. There he catches the last rays of the western light before he closes his book for the day.

CHAPTER X.

ST. DEINIOL'S HOSTEL.

An Old "Grammar School" and its New Utilisation—The Object of the Hostel—The Attractions of Hawarden—A Chance for Poor Students—A Tea Party at the Hostel.

STANDING outside St. Deiniol's Library door, and looking straight ahead, you have before you an old world entrance to a country house. It is perfectly simple, but picturesque. On one side a white stone wall, with ivy, and jessamine, and roses creeping all over it, and further on, where it joins the house, smothering that also in the sweet greenery. A bank of ferns and pinks and pansies runs along the wall and under the windows of the Hostel. On the other side of the drive leading to the Hostel, the lawn, bordered with rose bushes, now covered with flame-coloured haws, and with dwarf trees full of red-cheeked apples, runs to the low churchyard wall, which, however, is hidden behind a thick wreath of shrubs.

ST. DEINIOL'S HOSTEL.

In the background, overshadowed by old trees, is a stone wall, hidden under the ivy, and through a Gothic door in it you go into the churchyard, and to the Parish Church, St. Deiniol's, from which the Hostel and Library receive their names.

The Hostel itself, a square, two-storied brick house, can boast of no architectural attractions. But it is interesting, for all that, for it bears on its wall the date 1606, and has for close upon three centuries weathered many a storm. Up to quite recently, when the working of the Intermediate Education Act caused their eviction, the studious youth of Hawarden and the surrounding district occupied the house, which still locally goes by the name of the "Grammar School." But the rooms of the main building have already been converted into sitting-rooms, studies, or bedrooms, for those theologians or students who avail themselves of the opportunity offered to them by Mr. Gladstone for a quiet retreat among congenial books. The following leaflet, drawn out by Mr. Gladstone shortly before the Library was opened, is sufficient explanation of the objects of the Library and Hostel scheme. The leaflet, we should add, was sent to such of Mr. Gladstone's friends as are interested in the scheme :—

"Mr. Gladstone's Library.—St. Deiniol's Theological and General Library, close to Hawarden Church— soon to be put under a trust by Mr. Gladstone—opened about January 25th, 1894, for the use of students (lay and clerical, of any age), inquirers, authors, and clergy, or others desiring times of rest. Twenty-five thousand volumes already. A Hostel close by affords simple and

ST. DEINIOL'S HOSTEL: THE COMMON ROOM.

comfortable board and lodging—about twenty-five shillings weekly, or on special terms according to accommodation. One of the parish clergy will reside at the Hostel. Hawarden is high and healthy, commands good views, is easily accessible from many directions, and is on the main ways to the Welsh mountain scenery and watering places. Daily services and constant celebrations of Holy Communion in the Parish Church. Residence is invited either for a long or short time. Hawarden Village is the centre of the

ST. DEINIOL'S HOSTEL: THE DINING ROOM.

large Parish of Hawarden, and is reached as follows: (1) Hawarden Station (five minutes off) on line between Manchester (Cheshire Lines), the Northgate Station at Chester, and Wrexham; a direct line to Liverpool will open shortly; (2) Sandycroft or Queen's Ferry, on Chester and Holyhead Railway, about two miles;

(3) Broughton Hall, on Chester, Mole, and Denbigh Railway, about two and a half miles. A fly or sheet carriage at Glynne Arms, Hawarden. Apply at any time, with references if necessary, to the Rev. the Warden, St. Deiniol's Hostel, Hawarden, for further information."

We may here add that a fund has been started to enable men to come at a lower rate than twenty-five shillings a week, where, as is often the case, they cannot really afford it. Curates, for instance, are usually extremely poor. Anyone who likes to subscribe to this fund may feel that he is making good use of his money. The most ancient part of the house contains one large room, which, until recently, has been the scene of the "grammar-boys'" educational efforts. It is delightfully quaint, with its irregular windows, its heavy beams across the ceiling, and its general old world air. From one large Gothic window of this room, which is to be the common room of the Hostel, you look into the churchyard, through the network of trembling gold represented just now by the foliage of a rowan tree. Narrow, winding stairs, of solid black oak, slippery with age and worn out by the feet of many generations, lead from this part of the Hostel to some of the upper rooms. Altogether, the Hostel, with its nooks and corners, and cheerful, cosy rooms, seems just the right place for the

ST. DEINIOL'S HOSTEL: THE PRAYER ROOM.

abode of men in search of rest and peace and serious study.

On a winter afternoon, not long ago, we were the privileged guests at a tea party in the dining-room of St. Deiniol's. The firelight played in the early dusk over a very modern terra-cotta art paper, and the conversation was modern enough, and included such subjects as the latest of London stage plays (for the original Taffy of du Maurier's "Trilby" lives close by), the most sensational and realistic of new novels, the music of the present and the future, etc., etc. And the Warden of St. Deiniol's, Mr. Drew, who took the head of the table, and Mrs. Drew dispensing tea at the other end, and inmates and guests of the Hostel all joined cheerfully in this essentially modern tea-table talk. But through the sound of voices, and above the clatter of cups and spoons, you seemed, as twilight fell, to hear whisperings manifold, and were haunted by many a tale of times departed. For it is nearly three hundred years since the oldest part of St. Deiniol's Hostel was built, and the spirits of many generations haunt the old reposeful place.

CHAPTER XI.

MR. GLADSTONE'S HOME-LIFE.

His Country Clothes—Some Family Memorials—His Method of Reading—His Memory—A Glimpse of the Dining-room—Mr. Gladstone's Dietary—The Bedroom and its Pictures—Mr. Gladstone's Text.

Mr. Gladstone's home-life is characterised by its extreme simplicity and regularity. Being absorbed in great things does not mean, with him, a disregard of little things such as is affected by many of the would-be great. And although we cannot vouch for the accuracy of the story which every country practitioner feels bound to tell his dyspeptic patients, and according to which Mr. Gladstone makes "thirty-two bites" of every mouthful of food he eats, he certainly is careful to preserve his health and his working power by every sensible precaution. The old Statesman, with his fine, pale, gentle face, is an interesting figure as he walks lightly and briskly along the country road, silently acknowledging the fervent salutations of his friends the Hawarden villagers. He wears a coat, well-buttoned up, a long shawl wrapped closely round his neck, and a soft felt hat—a very different figure from that of the Prime Minister as he used to be known in London.

The Castle library, where Mr. Gladstone spends most of his time, is by no means a show place. You see at once that the books are frequently used, and that while Homer and Virgil, and Horace and Dante, have indeed their full share of attention, Maupassant and Marion Crawford—to choose only two of the innumerable array of modern writers—are not forgotten. Occasionally Mr. Gladstone hands over a batch of some two hundred novels to his daughter, Mrs. Drew, with the request to divide the sheep from the goats among them. That is to say, all the books are divided into three classes—novels worth keeping, novels to be given away, and novels to be destroyed.

Quite recently a most interesting collection of MSS. was heaped up on one of the tables in the library. They were divided into small bundles, neatly tied up, and carefully labelled in Mr. Gladstone's own handwriting. Here was a batch of "Confidential Letters on the Irish Church;" there another of "Verses at Eton, 1821 and after;" "Memorials of the Children," family letters, etc., etc. There are several busts in the room, and on the mantelpiece stands a small copy of Thorwaldsen's beautiful "Christ." Axes and sticks, and perchance a frankly disreputable looking umbrella, stand about in the corners; and across a quaint earthenware bowl lies an illuminated text: "Thy God hath sent forth strength for thee." The walls are covered with books, and volumes are also massed on large shelves jutting out from the walls into the room. Between each partition of books there is room to walk; thus the saving of space in arranging the library in this manner is enormous.

The members of Mr. Gladstone's family have seen it often, and for many years, but they have have not yet grown tired of watching him when, after an absence, he returns home to Hawarden, and enters again his beloved library. Nor can they speak without emotion of that look of perfect happiness and peace that beams from his eyes on such occasions. All the long morning he spends in the silence of that corner-room on the ground floor of Hawarden Castle.

There is not a single room at Hawarden where there is not an abundance of books. They are everywhere, and where other people have cosy corners and cabinets of nicknacks you have cases and shelves of books at Hawarden. They have penetrated, not in single copies or vellum-bound *éditions de luxe*, but in serried ranks into all rooms. They help to furnish the bright drawing-room, give an air of comfort to the anteroom with its enormous cases of valuable china, and are not absent even from Mr. Gladstone's bedroom and dressing-room. But in their thousands they are only to be seen in that ideal library where so many of his happiest days and years have been spent, ever since— nearly sixty years ago—Mr. Gladstone came first to Hawarden.

Mr. Gladstone has usually three books in reading at the same time, and changes from one to the other when his mind has reached the limit of absorption. This is a necessary corrective to the tendency to think only of one thing at one time, which was the source of much of his power, and of some weakness in politics. At one time he was so absorbed in the Eastern Question that he could hardly be induced to spare a thought for Ireland; later, it was just as difficult to get him to think of any political question but that of Ireland.

He complains sometimes that his memory is no longer quite so good as it used to be; but although this may be true, it is still twice as good as anybody else's, for Mr. Gladstone has an extraordinary faculty of not only remembering those things he ought to remember, but for forgetting those things it is useless for him to remember. His mind is thus unencumbered with any unnecessary top hamper, and he can always, so to speak, lay his hand upon anything the moment he wants it. This retentive memory was no doubt born with him, but it has been largely developed by the constant habit of taking pains. When he reads a book he does so pencil in hand, marking off on the margin those passages which he wishes to remember, querying those about which he is in doubt, and putting a cross opposite those which he disputes. At the end of the volume he constructs a kind of index of his own, which enables him to refer to those things he wishes to remember in the book.

Luncheon at Hawarden is pleasantly unconventional. The "lunch is on the hob" for several hours, to be partaken of when it suits the convenience of the various members of the household. The dining-room is one of the most cheerful and charming of all the rooms on the ground floor. It has a large oriel window, looking upon a quaint, old-world flower garden, and beyond this, through the trees, upon the ruins of old Hawarden Castle. Over the mantel-piece hangs a large oil painting of Mr. Gladstone as a young man, and portraits of his little granddaughter stand below it. On the opposite wall hangs a painting of Mrs. Gladstone as a young and very beautiful girl, There are plants and flowers everywhere; and near the window, adding to the air of homeliness, is a piano.

Mr. Gladstone is extremely simple in his tastes, and cheerfully allows the daintiest of things good to eat to pass by him, so long as his "outward conscience"— embodied for many years, and until death, by his friend and physician, the late Sir Andrew Clark—puts his veto on them. But quite apart from medical orders, he dislikes all rich and elaborate food. Light and simple dishes, blunt, kind-hearted Sir Andrew prescribed peremptorily, and sat down to write a most uninteresting dietary, every word and letter of which was strictly obeyed without a murmur. But though abstemious in all things gastronomic, Mr. Gladstone is not in any sense ascetic, but a believer in good cheer and plenty of it, for those who may partake of it without regret.

Mr. Gladstone's conversational gifts are enjoyed by his own family circle exclusively as often as by his visitors. On quiet evenings, when there are no visitors at Hawarden, he is perfectly contented to chat with his own friends, or sit quietly in his chair with a book for his companion. Anti-tobacconists will be glad to hear that he does not smoke, and never has smoked, belonging to the older school, which acquired its habits at a time when tobacco-smoking was regarded as somewhat vulgar. Hence neither pipe, cigar, nor cigarette is ever to be seen between his lips. Shortly after ten Mr. Gladstone's day is over, and he retires, for, on principle, he never allows himself to be cheated of sleep. "In the most exciting political crisis," he said on one occasion, "I dismiss current matters entirely from my mind when I go to bed, and will not think of them till I get up in the morning. I told Bright this, and he said: 'That's all very well for you, but my way is exactly the reverse. I think over all my speeches in bed.'" Seven hours' sleep is Mr. Gladstone's fixed allowance, "and," he added, with a smile, "I should like to have eight. I hate getting up in the morning, and hate it the same every morning. But one can do everything by habit, and when I have had my seven hours' sleep my habit is to get up." Mr. Gladstone's bedroom is on the first floor, which is reached by a fine staircase. It is a moderate-sized room, with two windows looking upon the turf, the bracken, and the winding paths of the grounds. The room is connected with Mr. Gladstone's dressing-room, which, for perfect simplicity, may be classed with the famous unpretentious rooms of other Grand Old Men. In the bedroom likewise, everything is perfectly simple and homely. On either side of the old-fashioned double bed the walls are covered with photographs of all sizes of Mr. and Mrs. Gladstone's children, and on the dressing-table, beside a smaller photograph of Mrs. Gladstone, a large portrait of Mr. John Morley has its place of honour.

In front of it is an illuminated text: "Thou wilt keep him in perfect peace whose mind is stayed on Thee." On that text every chapter of this book, describing Mr. Gladstone in the Evening of His Days, is a commentary.

www.ingramcontent.com/pod-product-compliance
Lightning Source LLC
Chambersburg PA
CBHW030303170426
43202CB00009B/850